AN INTRODUCTION TO CREDIT DERIVATIVES

SECOND EDITION

People still crave explanations even when there is no underlying understanding about what's going on ... erratic stock market movements always find a ready explanation in the next day's financial columns: a price rise is attributed to sentiment that 'pessimism about interest rate increases was exaggerated,' or to the view that 'company X had been oversold.' Of course these explanations are always *a posteriori*: commentators could offer an equally ready explanation if a stock had moved the other way.

Professor Martin Rees
Our Cosmic Habitat
London: Phoenix (2003), page 101

AN INTRODUCTION TO CREDIT DERIVATIVES

SECOND EDITION

Moorad Choudhry

AMSTERDAM • BOSTON • HEIDELBERG • LONDON
NEW YORK • OXFORD • PARIS • SAN DIEGO
SAN FRANCISCO • SINGAPORE • SYDNEY • TOKYO
Butterworth Heinemann is an imprint of Elsevier

Butterworth-Heinemann is an imprint of Elsevier
The Boulevard, Langford Lane, Kidlington, Oxford OX5 1GB, UK
225 Wyman Street, Waltham, MA 02451, USA

First edition published 1988
Second edition published 2013

© 2013, Moorad Choudhry and Elsevier Ltd. All rights reserved

The right of Moorad Choudhry to be identified as the author of this work has been asserted in accordance with the Copyright, Designs and Patents Act 1988

No part of this publication may be reproduced in any material form (including photocopying or storing in any medium by electronic means and whether or not transiently or incidentally to some other use of this publication) without the written permission of the copyright holder except in accordance with the provisions of the Copyright, Designs and Patents Act 1988 or under the terms of a licence issued by the Copyright Licensing Agency Ltd, 90 Tottenham Court Road, London, England W1T 4LP. Applications for the copyright holder's written permission to reproduce any part of this publication should be addressed to the publisher

Permissions may be sought directly from Elsevier's Science and Technology Rights Department in Oxford, UK: phone: (+44) (0) 1865 843830; fax: (+44) (0) 1865 853333; e-mail: hpermissions@elsevier.co.uk. You may also complete your request on-line via the Elsevier homepage (http://www.elsevier.com), by selecting 'Customer Support' and then Obtaining Permissions'

The views expressed in this book are an expression of the author's personal views only and do not necessarily reflect the views or policies of The Royal Bank of Scotland Group plc, its subsidiaries or affiliated companies, or its Board of Directors. RBS does not guarantee the accuracy of the data included in this book and accepts no responsibility for any consequence of their use. This book does not constitute an offer or a solicitation of an offer with respect to any particular investment.

Whilst every effort has been made to ensure accuracy, no responsibility for loss occasioned to any person acting or refraining from action as a result of any material in this book can be accepted by the author, publisher or any named person or corporate entity.

No part of this book constitutes investment advice and no part of this book should be construed as such. Neither the author nor the publisher or any named individual or entity is soliciting any action, response or trade in response to anything written in this book.

British Library Cataloguing in Publication Data
A catalogue record for this book is available from the British Library

Library of Congress Cataloguing in Publication Data
A catalogue record for this book is available from the Library of Congress

ISBN: 978-0-08-098295-3

For information on all Elsevier Butterworth-Heinemann finance publications visit our website at http://books.elsevier.com/finance

Typeset by MPS Limited, Chennai, India
www.adi-mps.com

**Working together to grow
libraries in developing countries**

www.elsevier.com | www.bookaid.org | www.sabre.org

ELSEVIER   BOOK AID International   Sabre Foundation

Dedication

Dedicated to my wife, Lindsay Choudhry

The street of shame is a street of hacks
The men behind the men who do deals behind our backs
The fourth estate is a house of hate
Media pimps with scant regards for facts
They nail their victims with a telephoto to the ground
Feed the people on scraps of evil
And their daily diet — a plateful of hateful.

Private greed, not public need
Life's geared to profit, money, wealth
They can't get enough of it.

And in the papers the same old story
Every picture sells a Tory

Money, profit, wealth,
They can't get enough of it,
Money, profit, wealth
Can't get enough of it....

> — Redskins, *A Plateful of Hateful* (King/Dean/Hewes)
> from *Neither Washington nor Moscow*, London Records 1986
> Reproduced with permission.

Contents

About the Author	xi
Foreword	xiii
Preface	xv
Preface to the First Edition	xvii

1	**Credit Risk**	**1**
	1.1 The Concept of Synthetic Investment	1
	1.2 Banks and Credit Risk Transfer	4
	1.3 Credit Risk and Credit Ratings	6
	1.3.1 Credit Risk	7
	1.3.2 Credit Ratings	10
	1.3.3 Understanding Credit Ratings	10
	1.4 Corporate Recovery Rates	12
	1.4.1 Recovery Rates	13
	1.4.2 Observation From Before 2008 Crash	14
2	**Credit Derivative Instruments: Part I**	**17**
	2.1 Credit Risk and Credit Derivatives	18
	2.1.1 Credit Events	20
	2.2 Credit Derivative Instruments	21
	2.2.1 Introduction	21
	2.2.2 Funded and Unfunded Contracts	22
	2.3 Credit Default Swaps	23
	2.3.1 Structure	23
	2.3.2 Basket Default Swaps	26
	2.3.3 Unwinding a CDS Position	27
	2.4 Asset Swaps	28
	2.4.1 Description	28
	2.4.2 Illustration Using Bloomberg	30
	2.5 Total Return Swaps	30
	2.6 Index CDS: The iTraxx Index	36
	2.7 Settlement	40
	2.7.1 Contract Settlement Options	40
	2.7.2 Market Requirements	42
	2.7.3 Cash Settlement Mechanics	43
	2.8 Risks in Credit Default Swaps	44
	2.8.1 Unintended Risks in Credit Default Swaps	44
	2.8.2 Extending Loan Maturity	44
	2.8.3 Risks of Synthetic Positions and Cash Positions Compared	45

viii   CONTENTS

    2.9 Impact of the 2007–2008 Financial Crash: New CDS Contracts
and the CDS 'Big Bang'   45
         2.9.1 The CDS 'Big Bang'   46
         2.9.2 CDS and Points Upfront   46
         2.9.3 Contract Changes   48
    References   49
    Appendices   49

3   Credit Derivative Instruments: Part II   53

    3.1 Credit-Linked Notes   53
         3.1.1 Description of CLNs   54
         3.1.2 Illustrations   56
    3.2 CLNs and Structured Products   61
         3.2.1 Simple Structure   61
         3.2.2 The First-to-Default Credit-Linked Note   62
    References   64

4   Credit Derivatives: Basic Applications   65

    4.1 Managing Credit Risk   65
    4.2 Credit Derivatives and Relative Value Trading   67
         4.2.1 Credit Selection   68
         4.2.2 Credit Pair Trade   68
         4.2.3 Basket Credit Structure Trade   69
    4.3 Bond Valuation from CDS Prices: Bloomberg Screen VCDS   71
    4.4 Relative Value Trading: Sovereign Names   71
         4.4.1 Rationale   73
         4.4.2 Example Trade Cash Flows: June 2009   73
    4.5 Applications of Total Return Swaps   75
         4.5.1 Capital Structure Arbitrage   76
         4.5.2 Synthetic Repo   77
         4.5.3 The TRS as Off-Balance Sheet Funding Tool   78
    4.6 Applications for Portfolio Managers   79
         4.6.1 Enhancing Portfolio Returns   83
         4.6.2 Reducing Credit Exposure   83
         4.6.3 Credit Switches and Zero-Cost Credit Exposure   84
         4.6.4 Exposure to Market Sectors   84
         4.6.5 Credit Spreads   85
    References   85

5   Credit Derivatives Pricing and Valuation   87

    5.1 Introduction   87
    5.2 Pricing Models   88
         5.2.1 Structural Models   88
         5.2.2 Reduced form Models   89
    5.3 Credit Spread Modelling   92

	5.4 Product Pricing Approach	94
	5.4.1 The Forward Credit Spread	94
	5.4.2 Asset Swaps Pricing	96
	5.4.3 Total Return Swap (TRS) Pricing	98
	5.5 Credit Curves	98
	References	100
6	**Credit Default Swap Pricing**	**101**
	6.1 Theoretical Pricing Approach	101
	6.2 Market Pricing Approach	104
	6.3 Credit Derivatives Pricing in Volatile Environments: 'Upfront + Constant Spread'	106
	6.4 Quick CDS Calculator	108
	References and Bibliography	109
	Appendices	110
7	**The Asset Swap–Credit Default Swap Basis**	**123**
	7.1 Asset Swap Pricing	123
	7.1.1 Basic Concept	123
	7.1.2 Asset-Swap Pricing Example	125
	7.1.3 Pricing Differentials	125
	7.2 The Basis as Market Indicator	127
	7.3 Analyzing the Basis Spread Measure	128
	7.3.1 ASW Spread	129
	7.3.2 Z-Spread	130
	7.3.3 Critique Of The Z-Spread	130
	7.3.4 Adjusted Z-spread	131
	7.3.5 Adjusted Basis Calculation	131
	7.3.6 Illustration	132
	7.4 Market Observations	133
	7.5 The iTraxx Index Basis	136
	7.6 Negative Basis Trade: Checking the Theoretical Bond Price	138
	References	139
Glossary		**141**
Index		**149**

About the Author

Moorad Choudhry is Head of Treasury, Corporate Banking Division, at The Royal Bank of Scotland. He is Visiting Professor at the Department of Mathematical Sciences, Brunel University, and Visiting Teaching Fellow at the Department of Management, Birkbeck, University of London.

Moorad is a Fellow of the Chartered Institute for Securities & Investment and a Fellow of the *ifs-School of Finance*. He is on the Editorial Board of the *Journal of Structured Finance* and *American Securitization*, and is a member of the Board of Directors of PRMIA.

Foreword

We are witnessing one of the most important episodes in financial history. The situation started with the crisis in the US financial system, is now challenging the European Union project and will hopefully finish with some lessons learned and a more robust and efficient financial system.

Arguably, the start of this crisis goes further back in time with years of excess leverage in the private sector ignited by low interest rates and prodigious financial innovations. The failure by some agents to identify the risks associated with such leverage and the interconnectivity of a global financial community has resulted in the difficult economic environment we are now experiencing. Governments had to step in and assume the excesses of the private sector to contain the effects of a disordered asset reduction. The challenge is now of course the deleveraging of the public sector itself.

At the heart of this financial innovation were credit derivatives. Traditionally the concept of credit was intimately linked to funding. One could not get exposure to a company without lending to it either through a bond or a loan. The introduction of credit derivatives was particularly important as it enabled the transfer of credit risk without funding or a relationship with the issuer of the underlying credit.

A new world of possibilities was immediately created. Financial intermediaries could shape the risks of their portfolios through credit default swaps (CDS), buying and selling credit protection. Investors could have access to synthetically created customized products that fulfilled their requirements at virtually any time.

Credit worthiness was now truly an asset class of its own and credit derivatives expanded its applications as they transferred credit risk to counterparty risk. For lenders it is not only important to buy protection to offset the threat of default by a borrower, but it is equally important to assess the credit quality of the counterparty from whom the protection is bought, as well as the collateral terms. For instance, the examples of AIG and Lehman have demonstrated the importance of introducing counterparty value adjustments (CVA).

Much has been written about the role of credit derivatives in the development of the actual economic situation. In fact, a great part of what has been written condemns credit derivatives. But as with many other innovations in history, it is the usage and not the tool that creates the problems.

As over-the-counter (OTC) instruments, credit derivatives practitioners have been involved through different industry bodies, mainly the International Swaps and Derivatives Association (ISDA), in strengthening the documentation and making the liquidation process of defaults transparent, responding to challenges as they arose. This has been a good start and sets solid ground, but is not enough. Challenges remain to be addressed, such the concentration of the industry within a handful of counterparties, and the migration to central clearing counterparties (CCPs) to mitigate counterparty risk.

Policy makers have been busy preparing new regulations with the aim of implementing them as soon as possible hoping to avoid the mistakes of the past. Dodd-Frank, the Volcker Rule, BIS3 and CRD4, MiFID, Sovency II, among others, contain very plausible measures for financial institutions and insurers, but as the drafting of these rules is evolving a clear and timely implementation would be desirable to have a robust regulatory framework in the future.

It is therefore important to understand the credit product and have all the relevant information to be able to diligently get involved with credit derivatives. In this valuable book you will be able to learn about the theory, insights, usages and implications of credit derivatives.

Finally, we have now the possibility of reflecting on previous financial crises to understand how they eventually came to an end, and we have a unique opportunity to make sure that new such episodes are not repeated. Better information and a responsible application of financial innovation should guarantee an efficient flow of capital to help the development of prosperous and sustainable growth.

Juan Blasco
Head of Credit Products
Wholesale Banking and Markets
Lloyds Bank Group

19 October 2012

Preface

Finance is a dry, unemotional subject. At least, it should be. Discussing any aspect of it should not be a cause for undue stress or argument. It isn't as if one is debating abortion rights or the Arab–Israeli peace process when one enters into a conversation on financial products. Any discourse in this field should be logical and objective, aimed at arriving at a sound solution that meets the needs of the relevant stakeholders.

Sadly, in the era following the banking crash of 2008, when the US and a number of European governments had to spend a considerable sum of taxpayers' money saving their banking sector, we encounter considerable subjective and emotional debate in finance. Some commentators have suggested that 'CDOs' caused the financial crash. Others, including some so-called 'gurus' best known for specific trades they may have undertaken over 20 years ago or who worked in banking over 30 years ago, have suggested that credit default swaps (CDS) caused the crash and should be banned to prevent the next crash.

This is utter nonsense. Financial instruments didn't cause the crash any more than tulips caused the crash of 1637 or Gordon Gekko caused the crash of 1987. A number of factors, working in concert, combined to produce a market correction and what all crashes have in common is speculation, mis-pricing, greed, applying incorrect or inaccurate assumptions and an inept understanding of the basic principles of finance. For tulips we can substitute sub-prime mortgages, for example, but on its own a financial instrument does not cause a crash. If used in a certain way certain products may make it easier for the effects of mis-pricing to be transferred more quickly across the market, but of itself it is financially illiterate to suggest that CDOs or CDS caused the crash. To do so only reflects subjective and emotional thinking.

This book isn't about the crash of 2008 or indeed any other crash. Rather, it is an introduction to a particular type of bank risk management product known as the credit derivative. The emergence of this product in the financial markets in the mid-1990s resulted in a repeat of the classic 'tail wagging the dog' scenario that one had observed in the 1980s, when interest-rate derivatives were introduced. Now the principal instrument used for valuation and analysis of the credit asset class is the credit default swap, just as interest-rate derivatives are now used as a benchmark for the interest-rate market, not the equivalent cash instruments.

In this book we introduce credit derivatives for the practitioner or graduate student in accessible fashion. The emphasis is on early understanding, and practical application. Those wishing to study the nuances of the product in greater detail may wish to consult the author's book *Structured Credit Products* (John Wiley & Sons, 2010).

Layout of the book
This book is organized into seven chapters, as follows:
Credit Risk
Credit Derivatives I
Credit Derivatives II
Credit Derivatives Applications
Pricing
CDS Pricing
The Asset Swap–Credit Default Swap Basis

Comments on the text are welcome and should be sent to me care of Elsevier. I hope you enjoy reading the book and that it is of value to you.

Moorad Choudhry
Surrey, England
May 2012

Preface to the First Edition

A key risk run by investors in bonds or loans is *credit risk*, the risk that the bond or loan issuer will default on the debt. To meet the need of investors to hedge this risk, the market uses *credit derivatives*. These are financial instruments originally introduced to protect banks and other institutions against losses arising from default. As such, they are instruments designed to lay off or take on credit risk. Since their inception, they have been used by banks, portfolio managers and corporate treasurers to enhance returns, to trade credit, for speculative purposes and as hedging instruments.

This book aims to provide an introduction to credit derivative instruments, and their uses, for beginners. The credit derivatives market has grown spectacularly, in a relatively short time, to become a key component of the capital markets and one which embraces a wide range of participants. They are a vital part of the corporate bond market, used for hedging purposes as well as for speculative trading purposes. Like the earlier generation interest-rate derivatives, such as interest-rate swaps and options, they are over-the-counter (OTC) instruments and as such very flexible; they can be tailor-made to suit individual customer requirements and used for a wide range of applications.

The principle behind credit derivatives is straightforward, as will be seen from the following text. And the flexibility of credit derivatives, arising from their status as OTC products, provides users with a number of advantages as they can be tailored to meet specific user requirements. The use of credit derivatives assists banks and other financial institutions with re-structuring their businesses, because they allow banks to repackage and parcel out credit risk while retaining assets on the balance sheet (when required), and thus maintain client relationships. As the instruments isolate certain aspects of credit risk from the underlying loan or bond, it becomes possible to separate the ownership of credit risk from the other features of ownership associated with the credit-risky assets in question; in other words, we can isolate credit as an asset in its own right. This flexibility has given rise to the market in structured credit products such as synthetic structured products, which exist in many varieties. This book reviews the main credit derivative instruments, including credit default swaps, total return swaps and credit-linked notes. We consider the instruments themselves, their application and pricing.

Moorad Choudhry
May 2004

CHAPTER 1

Credit Risk

The development of the credit derivatives market, and hence the subsequent introduction of structured credit products, was a response to the rising importance attached to credit risk management. For this reason, we believe it is worthwhile beginning this book with a look at credit risk, credit risk transfer and credit ratings from first principles.

1.1 THE CONCEPT OF SYNTHETIC INVESTMENT

If one stops to consider it, banks have been 'selling protection' on their customers ever since they began formally borrowing and lending amongst their customers in the Italian city-states during the Middle Ages. We describe how.

Cash Investment

A Bank lends 100 florins to Mr Borrower for a period of 5 years, who agrees to pay interest of C% per year each year until loan expiry, at which point he will return the original 100 florins. This is shown in Figure 1.1. We assume that Mr Borrower does not default on payment of both interest and principal during the term of this loan.

The net gain to Bank after the 5 years is the interest of C each year, which after 5 years will be 5C.

Synthetic Investment

A Bank sells protection on Mr Borrower to Mr Practitioner for a period of 5 years, who agrees to pay C basis points premium per year each year until trade expiry. This is also shown in Figure 1.1. We assume that Mr Borrower does not go into bankruptcy, liquidation or administration during the 5 years of the contract between Bank and

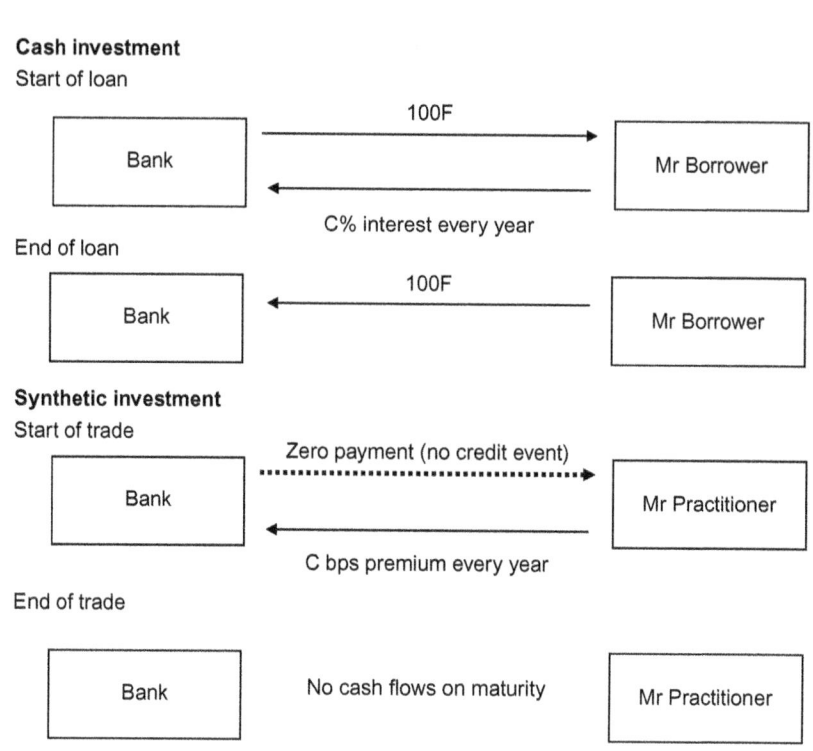

FIGURE 1.1 Cash and synthetic investment: no default.

Mr Practitioner, and that Mr Practitioner keeps up the premium payments until contract expiry.

The net gain to Bank after the 5 years is the credit default swap (CDS) premium of C each year, which after 5 years will be 5C. Note that Mr Borrower is not involved in this transaction at all, as we see in Figure 1.1. The contract is between Bank and Mr Practitioner.

This illustration is a bit cheeky but it makes the point. The return for Bank of C is identical in each case. So Bank can decide between cash and synthetic investment; in theory the return will be identical. In terms of net cash flow, the end-result actually is identical if the price associated with Mr Borrower credit risk is C. The only difference is that the cash trade is funded, while the CDS trade is unfunded. Bank has to find 100 florins to lend to Mr Borrower, but it doesn't need to find any cash to invest in Mr Borrower by means of the contract with Mr Practitioner.

Of course, if there is a credit event, then under the synthetic investment Bank will have to find 100 florins. So we need to consider the event of default. Before we do, let's make the illustration more real-world.

1.1 THE CONCEPT OF SYNTHETIC INVESTMENT

Real-World Application

If we convert the bond investment to an asset-swap that pays Libor + X basis points, the interest basis of the bond has changed from fixed coupon to floating coupon. It is now conceptually identical to the synthetic (credit default swap) premium of X basis points.

In the cash market, we assume that the investor has funding cost of Libor-flat, so that the 100 florins it borrows to buy the bond costs it Libor-flat in funding. The net return to the investor is X. For the synthetic investor, who has no funding cost (it does not borrow any cash as it is not buying a bond), the net return is X basis points.

Let us now allow for a default or 'credit event'.

Cash Investment

At the beginning of year 3, Mr Borrower is declared bankrupt and defaults on his debt to Bank. The Bank falls in line with the other creditors, and after 3 years receives a payout from the administrators of 39 cents on the florin.

After 5 years, Bank has therefore received 2C in interest, but lost 61 florins of capital, a substantial loss.

Synthetic Investment

At the beginning of year 3, Mr Borrower is declared bankrupt and immediately Bank pays 100 florins to Mr Practitioner. In return, Mr Practitioner delivers to Bank 100 florins nominal of a loan that Mr Borrower took out from A N Other Bank. This loan has some residual value to Bank, at the time it is valued at 30 cents on the florin so Bank records a capital loss of 70 florins, a substantial loss.

After 3 years the Mr Borrower loan that Bank is holding is valued at 39 cents on the florin, so Bank recovers 9 florins of the loss it booked 3 years ago. So after 5 years Bank has received 2C in interest, but lost 61 florins of capital, a substantial loss. The illustrations are shown in Figure 1.2.

The point we are making is that funded and unfunded investments are identical on a net–net basis, at least in theory, and involve taking credit risk exposure in Mr Borrower's name. The price of this risk is C%, and reflects what the market thinks of the credit quality of Mr Borrower. Of course there are some technical differences, not least the introduction of another counterparty, Mr Practitioner. But from the point of view of Bank, both trades are in essence identical, or at least have identical objectives. And the synthetic trade actually has some advantages, principally with regard to the fact that no funding is required. The counterparty risk to Mr Practitioner (principally his ability to keep up premium payments) can be viewed as a disadvantage.

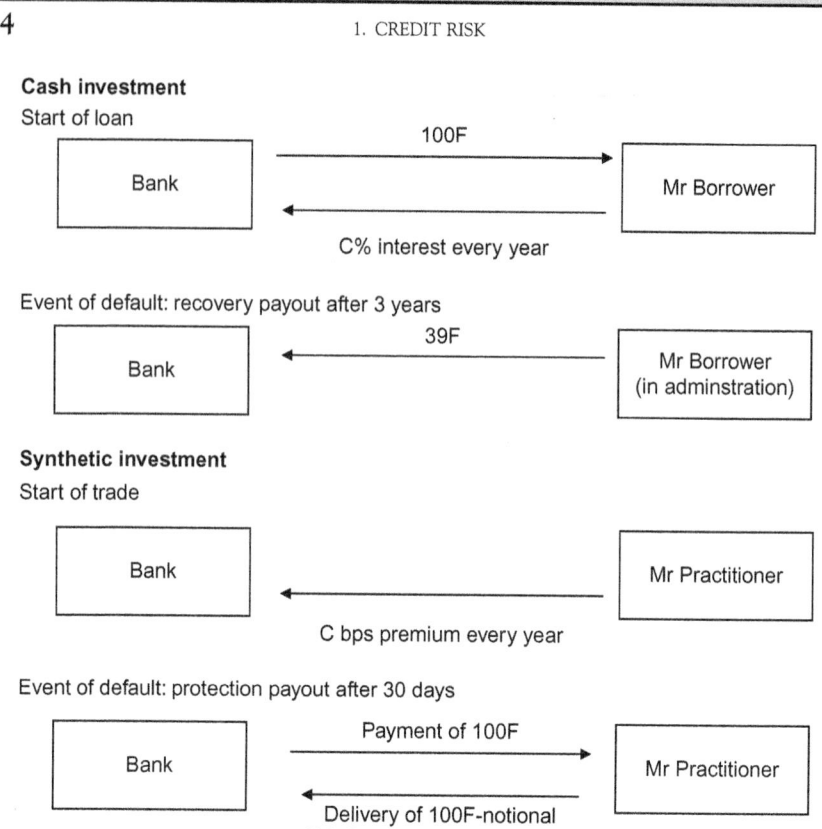

FIGURE 1.2 Cash and synthetic investment: default.

1.2 BANKS AND CREDIT RISK TRANSFER

Banking institutions have always sought ways of transferring the credit (default) loss risk of their loan portfolios, for two reasons: (i) to remove the risk of expected losses from their balance sheet, due to an expected increase in incidence of loan default and (ii) to free up capital, which can then be used to support further asset growth (increased lending). The first method of reducing credit risk is selling loans outright. This simply removes the asset from the balance sheet, and is in effect a termination of the transaction. Alternatives to this approach include the following:

- spreading the risk in the first place via a syndicated loan, in partnership with other banks;
- securitizing the loan, thus removing the asset off the balance sheet and, depending on how the transaction is structured and sold, transferring the credit risk associated with the loan;

1.2 BANKS AND CREDIT RISK TRANSFER

- covering the risk of default loss with a CDS or an index credit default swap;
- transferring the risk of the asset to a specialty finance company.

In principle, the ability to transfer credit risk is an advantage to the financial market as a whole, as it means risks can be held more or less evenly across the system. In practice this may not happen, and risk can end up being concentrated in particular sectors or amongst particular groups of investors. But such a result is not to detract from the inherent positive impact of credit risk transfer.

Credit risk transfer enables banks to manage bank capital more efficiently, and also to diversify their risk exposures. In theory this means they can reduce their overall credit risk exposure. (We emphasize 'in theory'. In practice, during a market downturn such as the financial crash of 2007–2008, diversified investment portfolios may perform no better than concentrated portfolios.)

The risk is transferred — not eliminated — to other banks and other market sectors such as institutional investors, hedge funds, corporations and local authorities. More efficient allocation of capital should, in theory, lead to a lower cost of credit, which is a positive development for the economy as a whole. Put against this is the negative notion that securitization and the use of CDS results in the poorest-quality assets always being retained on bank balance sheets, because such assets are illiquid and not attractive to other investors; this concentrates risk. In addition, loan risk may be transferred to investors who are less familiar with the end-obligor, and so are less able to monitor the borrower's performance and notice any deterioration in credit quality. Where banks originate debt that they then instantly transfer, they may pay less attention to borrower quality and repayment ability: this reduction in lending standards was one of the contributing factors to the US sub-prime mortgage market default.

In essence though, these are 'micro-level' arguments against, and we will proceed with the premise that credit risk transfer is on balance a beneficial activity.

Risk transfer is a logical part of bank risk management. The benefits of credit risk transfer are (i) reduction of capital costs that were associated with the full-risk-exposure loan asset and (ii) diversification of risk. Risk transfer via securitization produces other potential benefits, in the form of new investment product for bank and non-bank investors that is a liquid financial instrument, which may not have been the case for the securitized asset.

Risk transfer via CDS enables the bank to maintain the loan and thus the client relationship, while benefiting from a capital cost reduction. That the CDS can be executed with an identical maturity to the loan is one of the prime advantages of the CDS instrument.

Given that there are various drivers behind why a bank may select to transfer the credit risk of an asset on its balance sheet using a credit derivative, by extension one can see that the same advantages apply if a bank wishes to transfer entire books of risk via a basket credit derivative, which can be either a cash collateralized debt obligation (CDO) or synthetic CDO. From the other side, investor demand for the product that arises from a CDO deal is also a driver of the transaction. Synthertic CDOs are discussed in the author's book *Structured Credit Products*, 2nd edn (John Wiley & Sons, 2010).

1.3 CREDIT RISK AND CREDIT RATINGS

Credit derivatives are bilateral financial contracts that transfer credit default risk from one counterparty to another. They represent a natural extension of fixed income derivatives in that they isolate and separate the element of credit risk (arguably the largest part of a bank's risk profile) from other risks, such as market and operational risks. They exist in a variety of forms; perhaps the simplest is the credit default swap, which is conceptually similar to an insurance policy taken out against the default of a bond, for which the purchaser of the insurance pays a regular premium. However, credit derivatives are different from other forms of credit protection, such as guarantees and mortgage indemnity insurance, because:

- the borrower is generally asked for a mortgage indemnity policy, or a guarantee;
- the credit derivative is requested by the lending bank, and the borrower doesn't have to know that the transaction has taken place;
- credit derivatives are tradable, while other forms of protection are generally not.

The currency and bond market volatility in Asia in 1997 and 1998 demonstrated the value of credit derivatives. For example, in 1998 the International Finance Corporation of Thailand bought back $500 m of bonds several years before maturity because of a graduated put provision that was exercisable if the bank's credit rating fell below investment grade; the bond would have paid out an additional 50 basis points of yield if the bond fell two levels in creditworthiness and 25 basis points per additional level until the put threshold below investment grade was reached. This in fact occurred when Moody's re-rated Thailand to Ba1 grade. In volatile markets, investors are generally happy to give up yield in return for lower credit risk. Thus financial institutions have started focusing on credit as a separate asset class rather than treating counterparty credit risk as one of the risks associated with an asset.

1.3.1 Credit Risk

There are two main types of credit risk that a portfolio of assets, or a position in a single asset, is exposed to. These are credit default risk and credit spread risk.

Credit Default Risk

This is the risk that an issuer of debt (obligor) is unable to meet its financial obligations. This is known as *default*. There is also the case of technical default, which is used to describe a company that has not honoured its interest payments on a loan for (typically) 3 months or more, but has not reached a stage of bankruptcy or administration. Where an obligor defaults, a lender generally incurs a loss equal to the amount owed by the obligor less any recovery amount which the firm recovers as a result of foreclosure, liquidation or restructuring of the defaulted obligor. This recovery amount is usually expressed as a percentage of the total amount, and is known as the *recovery rate*. All portfolios with credit exposure exhibit credit default risk.

The measure of a firm's credit default risk is given by its *credit rating*. The three largest credit rating agencies are Moody's, Standard & Poor's and Fitch. These institutions undertake qualitative and quantitative analysis of borrowers, and formally rate the borrower after their analysis. The issues considered in the analysis include:

- the financial position of the firm itself, for example its balance sheet position and anticipated cash flows and revenues;
- other firm-specific issues, such as the quality of management and succession planning;
- an assessment of the firm's ability to meet scheduled interest and principal payments, both in its domestic and in foreign currencies;
- the outlook for the industry as a whole, and competition within it, together with general assessments of the domestic economy.

The range of credit ratings awarded by the three largest rating agencies is shown in Table 1.1. We will discuss credit ratings again shortly. Figure 1.3 shows these ratings on the Bloomberg RATD screen.

Credit Spread Risk

Credit spread is the excess premium, over and above government or risk-free risk, required by the market for taking on a certain assumed credit exposure. For instance, Figure 1.4 shows the credit spreads in April 2012 for US dollar corporate bonds with different credit ratings (AAA, A and BBB). The benchmark is the on-the-run or *active* US Treasury issue for the given maturity. Note that the higher the credit rating, the smaller the credit spread. Credit spread risk is the risk of

TABLE 1.1 Corporate Bond Credit Ratings

Fitch	Moody's	S&P	Summary Description
INVESTMENT GRADE – HIGH CREDITWORTHINESS			
AAA	Aaa	AAA	Gilt-edged, prime, maximum safety, lowest risk
AA +	Aa1	AA +	
AA	Aa2	AA	High-grade, high credit quality
AA –	Aa3	AA –	
A +	A1	A +	
A	A2	A	Upper-medium grade
A –	A3	A –	
BBB +	Baa1	BBB +	
BBB	Baa2	BBB	Lower-medium grade
BBB –	Baa3	BBB –	
SPECULATIVE – LOWER CREDITWORTHINESS			
BB +	Ba1	BB +	
BB	Ba2	BB	Low-grade; speculative
BB –	Ba3	BB –	
B +	B1		
B	B	B	Highly speculative
B –	B3		
PREDOMINANTLY SPECULATIVE, SUBSTANTIAL RISK OR IN DEFAULT			
CCC +		CCC +	
CCC	Caa	CCC	Substantial risk, in poor standing
CC	Ca	CC	May be in default, very speculative
C	C	C	Extremely speculative
		CI	Income bonds – no interest being paid
DDD			
DD			Default
D		D	

financial loss resulting from changes in the level of credit spreads used in the marking-to-market of a product. It is exhibited by a portfolio for which the credit spread is traded and marked. Changes in observed credit spreads affect the value of the portfolio and can lead to losses for investors.

1.3 CREDIT RISK AND CREDIT RATINGS

LONG-TERM RATING SCALES COMPARISON

MOODY'S	Aaa	Aa1	Aa2	Aa3	A1	A2	A3	Baa1	Baa2	Baa3
S&P	AAA	AA+	AA	AA-	A+	A	A-	BBB+	BBB	BBB-
COMP	AAA	AA+	AA	AA-	A+	A	A-	BBB+	BBB	BBB-
FITCH	AAA	AA+	AA	AA-	A+	A	A-	BBB+	BBB	BBB-
DBRS	AAA	AAH	AA	AAL	AH	A	AL	BBBH	BBB	BBBL
R&I	AAA	AA+	AA	AA-	A+	A	A-	BBB+	BBB	BBB-
JCR	AAA	AA+	AA	AA-	A+	A	A-	BBB+	BBB	BBB-
AM Best	aaa	aa+	aa	aa-	a+	a	a-	bbb+	bbb	bbb-

Note: white = investment grade, yellow = non-investment grade

FIGURE 1.3 Bloomberg screen RATD, long-term credit ratings. © *Bloomberg L.P. Reproduced with permission.*

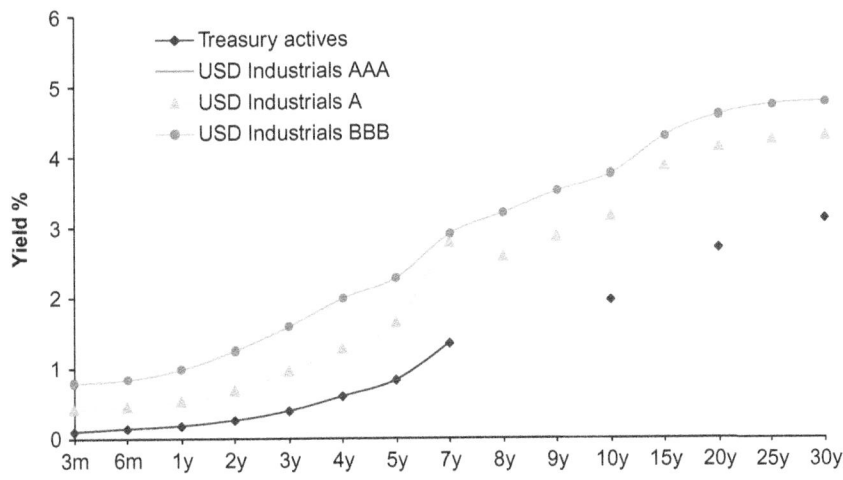

FIGURE 1.4 US dollar bond yield curves, April 2012. *Source: Bloomberg.*

1.3.2 Credit Ratings

The risks associated with holding a fixed interest debt instrument are closely connected with the ability of the issuer to maintain the regular coupon payments as well as redeem the debt on maturity. Essentially, the *credit risk* is the main risk of holding a bond. Only the highest quality government debt, and a small number of supra-national and corporate debt, may be considered to be entirely free of credit risk. Therefore, at any time the yield on a bond reflects investors' views on the ability of the issuer to meet its liabilities as set out in the bond's terms and conditions. A delay in paying a cash liability as it becomes due is known as technical default, and is a cause for concern for investors; failure to pay will result in the matter being placed in the hands of the legal court as investors seek to recover their funds. To judge the ability of an issue to meet its obligations for a particular debt issue, for the entire life of the issue, requires judgemental analysis of the issuer's financial strength and business prospects. There are a number of factors that must be considered, and larger banks, fund managers and corporates carry out their own *credit analysis* of individual borrowers' bond issues. The market also makes a considerable use of formal *credit ratings* that are assigned to individual bond issues by a formal credit rating agency.

The specific factors that are considered and the methodology used in conducting the analysis differs slightly amongst the individual ratings agencies. Although in many cases the ratings assigned to a particular issue by different agencies are the same, they occasionally differ and in these instances investors usually seek to determine what aspect of an issuer is given more weight in an analysis by which individual agency. Note that a credit rating is not a recommendation to buy (or equally sell) a particular bond, nor is it a comment on market expectations. Credit analysis does take into account general market and economic conditions, but the overall point of credit analysis is to consider the financial health of the issuer and its ability to meet the obligations of the specific issue being rated. Credit ratings play a large part in the decision-making of investors, and also have a significant impact on the interest rates payable by borrowers.

1.3.3 Understanding Credit Ratings

A credit rating is a formal opinion, given by a rating agency, of the *credit risk* for investors in a particular issue of debt securities. Ratings are given to public issues of debt securities by any type of entity, including governments, banks and corporates. They are also given to short-term debt such as commercial paper, as well as bonds and medium-term notes.

Purpose of Credit Ratings

Investors in securities accept the risk that the issuer will default on coupon payments or fail to repay the principal in full on the maturity date. Generally credit risk is greater for securities with a long maturity, as there is a longer period for the issuer potentially to default. For example, if a company issues 10-year bonds, investors cannot be certain that the company will still exist in 10 years' time: it may have failed and gone into liquidation some time before that. That said, there is also risk attached to short-dated debt securities; indeed, there have been instances of default by issuers of commercial paper, which is a very short-term instrument.

The prospectus or offer document for an issue provides investors with some information about the issuer so that some credit analysis can be performed on the issuer before the bonds are placed. The information in the offer documents enables investors themselves to perform their own credit analysis by studying this information before deciding whether or not to invest. Credit assessments take time, however, and also require the specialist skills of credit analysts. Large institutional investors do in fact employ such specialists to carry out credit analysis; however, often it is too costly and time-consuming to assess every issuer in every debt market. Therefore, investors commonly employ two other methods when making a decision on the credit risk of debt securities:

- name recognition;
- formal credit ratings.

Name recognition is when the investor relies on the good name and reputation of the issuer and accepts that the issuer is of such good financial standing, or is of sufficient financial standing, that a default on interest and principal payments is highly unlikely. An investor may feel this way about, say, Microsoft or British Petroleum plc.

Formal Credit Ratings

Credit ratings are provided by the specialist agencies. The major credit rating agencies are Standard & Poor's, Fitch and Moody's, based in the United States. There are other agencies both in the US and other countries. On receipt of a formal request, the credit rating agencies will carry out a rating exercise on a specific issue of debt capital. The request for a rating comes from the organization planning the issue of bonds. Although ratings are provided for the benefit of investors, the issuer must bear the cost. However, it is in the issuer's interest to request a rating as it raises the profile of the bonds, and investors may refuse to buy paper that is not accompanied by a recognized rating.

Although the rating exercise involves a credit analysis of the issuer, the rating is applied to a specific debt issue. This means that in theory the credit rating is applied not to an organization itself, but to specific debt securities that the organization has issued or is planning to issue. In practice, it is common for the market to refer to the creditworthiness of organizations themselves in terms of the rating of their debt.

The rating for an issue is kept constantly under review, and if the credit quality of the issuer declines or improves the rating will be changed accordingly. An agency may announce in advance that it is reviewing a particular credit rating, and may go further and state that the review is a precursor to a possible downgrade or upgrade. This announcement is referred to as putting the issue under *credit watch*. The outcome of a credit watch is in most cases likely to be a rating downgrade; however, the review may re-affirm the current rating or possibly upgrade it. During the credit watch phase, the agency will advise investors to use the current rating with caution. When an agency announces that an issue is under credit watch, the price of the bonds will fall in the market as investors look to sell out of their holdings. This upward movement in yield will be more pronounced if an actual downgrade results. For example, in 2010 the government of Spain was placed under credit watch and subsequently lost its AAA credit rating; as a result there was an immediate and sharp sell-off in Spanish government Eurobonds before the rating agencies had announced the actual results of their credit review.

1.4 CORPORATE RECOVERY RATES

When a corporate obligor experiences bankruptcy or enters into liquidation or administration, it will default on its loans. However, this does not mean that all the firm's creditors will lose everything. At the end of the administration process, the firm's creditors typically will receive back a portion of their outstanding loans, a *recovery* amount.[1] The percentage of the original loan that is received back is known as the *recovery rate*, which is defined as the percentage of par value that is returned to the creditor.

The seniority of a loan strongly influences the level of the recovery rate. Table 1.2 shows recovery rates for varying levels of loan seniority in 2008 as published by Moody's. The standard deviation for each recovery rate reported is high, which illustrates the dispersion around the mean and reflects widely varying recovery rates even within the

---

[1] This recovery may be received in the form of other assets, such as securities or physical plant, instead of cash.

TABLE 1.2 Moody's Recovery Rates, According to Loan Seniority, for 2008. Reproduced with permission

Seniority	Mean	Standard Deviation
Senior secured bank loans	60.70%	26.31%
Senior secured	53.83%	25.41%
Senior unsecured	52.13%	25.12%
Senior subordinated	39.45%	24.79%
Subordinated	33.81%	21.25%
Junior subordinated	18.51%	11.26%
Preference shares	8.26%	10.45%

same level of seniority. It is clear that the more senior a loan or a bond, the higher recovery it will enjoy in the event of default.

1.4.1 Recovery Rates

As we shall see in Chapter 5, the concept of the 'recovery rate' (RR) is a key parameter in CDS valuation. This is somewhat unfortunate, because the nature of markets is such that an assumed rate must be used. In the real world, actual recovery value from a defaulted obligation may not be known for some years.

The procedure for determining the recovery rate in the cash market is a long drawn-out affair. Debt investors take their place in the queue behind all other creditors and receive their due after the administrators have completed their work with. This process can take a matter of months or over 10 years. The rating agencies make an assumption of what the final recovery amount will be from the market price of the debt asset at the time bankruptcy or default is announced. This approach is carried over to an extent into the credit derivative market.

The definition of recovery rate in the CDS market differs slightly from that in the cash bond market, for reasons of practicality. This is because the contract must settle 30 days after the notice of a credit event has been announced, and the real 'recovery rate' is not known by then. At the same time, the model approach under which the CDS would have been priced and valued up to now would have used a 'recovery rate' as one of its parameters. So in the CDS market 'recovery' is defined as the market value of the 'delivered obligation'. This market value is determined by a poll of CDS dealers bidding for the defaulted assets of the reference entity.

Note that recovery rates for the synthetic market will therefore differ from those in the cash market in practice. This arises for a number of reasons, one of which is that in the CDS market a 'credit event' will encompass circumstances that fall short of full default in the cash market. For example Moody's notes three categories of default for the purposes of its ratings and historical default statistics:

- delayed or failed coupon or principal payments;
- bankruptcy or receivership;
- distressed exchange that results in investors having a lower obligation value, undertaken by the obligor in order to avoid default.

1.4.2 Observation From Before 2008 Crash

Recovery rates in practice vary widely, as we note below. In essence, what the 2007–2008 credit crunch has taught us is that if one is using CDS to hedge credit risk, the safest approach is to assume a 0% recovery rate. This is because in the event of default it will be some time before the investor receives the recovery value, while in the meantime the payment from the CDS that was used to hedge the asset will be (100 − RR) so the investor will actually have lost out on some of the investment until recovery is received. In the meantime the investor's accountants would most probably have written down the entire par

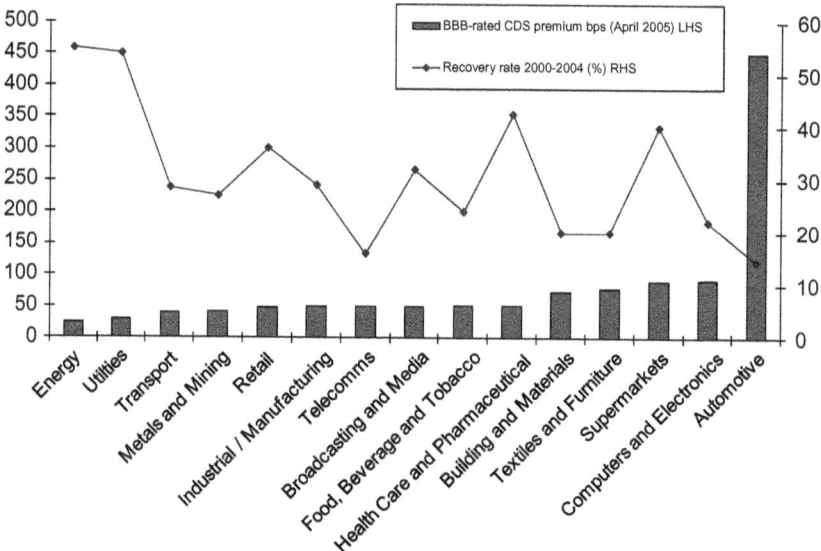

FIGURE 1.5 BBB CDS level versus historic recovery rate. Rates are average for industry in April 2005. *Source: Fitch. Reproduced with permission.*

1.4 CORPORATE RECOVERY RATES

value of the investment. Therefore, for complete hedging it is best to assume a zero recovery value when calculating the notional amount of the CDS used to hedge the investment.

Despite the RR being a key parameter in CDS pricing models, it is not apparent that it is influencing CDS premiums heavily, or that actual historical RRs are used when selecting the input level. A look at CDS spreads during the period 2000–2004 shows that while 'recovery value' varies widely, the CDS premium across all industries is fairly similar (see Figure 1.5)

We observe that most industries were trading at levels fairly close to each other: within 10 or 20 bps. Industries traditionally regarded as higher risk (e.g. computers and electronics) or operating in more difficult circumstances, such as automobile, are marked up. But otherwise the closeness is noteworthy.

CHAPTER 2

Credit Derivative Instruments
Part I

In Chapter 1 we considered the concept of credit risk and credit ratings. Credit derivatives, introduced in 1993, isolate credit as a distinct asset class, much like how interest-rate derivatives, such as swaps and futures, isolated interest rates in the 1980s. This isolation of credit has improved the efficiency of the capital markets, because market participants can now separate the functions of credit origination and credit-risk bearing. Banks have been able to spread their credit risk exposure across the financial system, which arguably reduces *systemic* risk. They also improve market transparency by making it possible to price specific types of credit risk better.[1] In this chapter, we consider the various *unfunded* credit derivative instruments.[2] We will go on later to look at various applications of the instruments and their pricing and valuation. We begin with some observations on market participants and applications.

---

[1] Some commentators have suggested that credit derivatives have *reduced* market transparency because it may not be possible to track where credit risk has gone after it has been removed from bank balance sheets. It has also been suggested that use of credit derivatives increases systemic risk because they spread risk and hence increase the risk of contagion. This debate is ongoing.

[2] This term is explained in this chapter. The next chapter looks at *funded* credit derivatives.

2.1 CREDIT RISK AND CREDIT DERIVATIVES

Credit derivatives are financial contracts designed to reduce or eliminate credit risk exposure by providing insurance against losses suffered due to *credit events*. A payout under a credit derivative is triggered by a credit event associated with the credit derivative's *reference asset* or *reference entity*. As banks define default in different ways, the terms under which a credit derivative is executed usually include a specification of what constitutes a credit event. The principle behind credit derivatives is straightforward. Investors desire exposure to non-default-free debt because of the higher returns this offers. However, such exposure brings with it concomitant credit risk. This can be managed with credit derivatives. Alternatively, the credit exposure itself can be taken on synthetically if, for instance, there are compelling reasons why a cash market position cannot be established. The flexibility of credit derivatives provides users with a number of advantages, and as they are over-the-counter (OTC) products they can be designed to meet specific user requirements.

In Chapter 4 we will look further at the main applications for which credit derivatives are used. Some of the most common reasons for which they are used include:

- hedging credit risk (this includes credit default risk, dynamic credit risks and changes in credit quality);
- reducing credit risk with a specific client (*obligor*) so that lending lines to this client are freed up for further business;
- diversifying investment options.

We focus on credit derivatives as instruments that may be used to manage risk exposure inherent in a corporate or non-AAA sovereign bond portfolio, and to manage the credit risk of commercial bank loan books. The intense competition amongst commercial banks, combined with rapid disintermediation, has meant that banks have been forced to evaluate their lending policy with a view to improving profitability and return on capital. The use of credit derivatives assists banks with restructuring their businesses, because they allow banks to repackage and parcel out credit risk while retaining assets on balance sheet (when required) and thus maintain client relationships. As the instruments isolate credit risk from the underlying loan or bond and transfer them to another entity, it becomes possible to separate the ownership and management of credit risk from the other features of ownership associated with the assets in question, such as customer franchise. This means that illiquid assets such as bank loans, and illiquid bonds, can have their credit risk exposures transferred; the bank owning the assets can protect against credit loss even if it cannot transfer the assets themselves.[3]

The same principles carry over to the credit risk exposures of portfolio managers. For fixed-income portfolio managers, some of the advantages of using credit derivatives include the following:

- They can be tailor-made to meet the specific requirements of the entity buying the risk protection, as opposed to the liquidity or term of the underlying reference asset.
- They can be 'sold short' without risk of a liquidity or delivery squeeze, as it is a specific credit risk that is being traded. In the cash market it is not possible to 'sell short' a bank loan, for example, but a credit derivative can be used to establish synthetically the economic effect of such a position.
- As they theoretically isolate credit risk from other factors such as client relationships and interest-rate risk, credit derivatives introduce a formal pricing mechanism to price credit issues only. This means a market is available in credit only, allowing more efficient pricing, and it becomes possible to model a term structure of credit rates.
- They are off-balance-sheet instruments,[4] and as such incorporate a certain flexibility and leverage, exactly like other financial derivatives. For instance, bank loans are not particularly attractive investments for certain investors because of the administration required in managing and servicing a loan portfolio. However, an exposure to bank loans and their associated return can be achieved using credit derivatives while simultaneously avoiding the administrative costs of actually owning the assets. Hence credit derivatives allow investors access to specific credits while allowing banks access to further distribution for bank loan credit risk.

Thus credit derivatives can be an important instrument for bond portfolio managers, as well as commercial banks, who wish to increase the liquidity of their portfolios, gain from the relative value arising from credit pricing anomalies, and enhance portfolio returns.

---

[3] As we note, the bank may not wish to transfer the assets, to maintain client relationships. It can also transfer the assets in a securitization transaction, which can also bring in funding (cash securitization).

[4] When credit derivatives are embedded in certain fixed-income products, such as structured notes and credit-linked notes, they are then off-balance-sheet but part of a structure that will have on-balance-sheet elements. Funded credit derivatives are on-balance-sheet.

2.1.1 Credit Events

The occurrence of a specified credit event will trigger the termination of the credit derivative contract, and transfer of the default payment from the protection seller to the protection buyer. The following may be specified as credit events in the legal documentation between counterparties:

- a downgrade in S&P and/or Moody's credit rating below a specified minimum level;
- financial or debt restructuring, for example occasioned under administration or as required under US bankruptcy protection;
- bankruptcy or insolvency of the reference asset obligor;
- default on payment obligations such as bond coupon and continued non-payment after a specified time period;
- technical default, for example the non-payment of interest or coupon when it falls due;
- a change in credit spread payable by the obligor above a specified maximum level.

The International Swaps and Derivatives Association (ISDA) has compiled standard documentation governing the legal treatment of credit derivative contracts. The standardization of legal documentation promoted the ease of execution and was a factor in the rapid growth of the market. The 1999 ISDA credit default swap documentation specified bankruptcy, failure to pay, obligation default, debt moratorium and restructuring to be credit events. Note that it does not specify a rating downgrade to be a credit event.[5]

A summary of the credit events as set forth in the ISDA definitions is given in Appendix 2.1.

The precise definition of 'restructuring' is open to debate, and has resulted in legal disputes between protection buyers and sellers. Prior to issuing its 1999 definitions, ISDA had specified restructuring as an event or events that resulted in making the terms of the reference obligation 'materially less favourable' to the creditor (or protection seller) from an economic perspective. This definition is open to more than one interpretation, and caused controversy when determining if a credit event had occurred. The 2001 definitions specified more precise conditions, including any action that resulted in a reduction in the amount of principal. In the European market, restructuring is generally retained as a credit event in contract documentation, but in the US market it is less common to see it included. Instead, US contract documentation tends to include as a credit event a form of *modified restructuring*, the impact of

---

[5]The ISDA definitions from 1999, the restructuring supplement from 2001 and the 2003 and 2009 definitions are available at www.ISDA.org.

which is to limit the options available to the protection buyer as to the type of assets it could deliver in a physically settled contract. Further clarification is provided in the 2003 ISDA definitions.[6]

2.2 CREDIT DERIVATIVE INSTRUMENTS

Before looking at the main types of credit derivative, we consider some generic details of all credit derivatives.

2.2.1 Introduction

Credit derivative instruments enable participants in the financial market to trade in credit as an asset, as they isolate and transfer credit risk. They also enable the market to separate funding considerations from credit risk. A number of instruments come under the category of credit derivatives, and in this and the next chapter we consider the most commonly encountered of these. Irrespective of the particular instrument under consideration, all credit derivatives can be described by the following characteristics:

- the *reference entity*, which is the asset or name on which credit protection is being bought and sold;[7]
- the credit event (or events) that indicate that the reference entity is experiencing or about to experience financial difficulty and act as trigger events for termination of and payments under the credit derivative contract;

[6]The debate on restructuring as a credit event arose out of a number of events, notably the case involving a corporate entity, Conseco, in the USA in 2000. It concerned the delivery option afforded the protection buyer in a physically settled credit derivative, and the *cheapest-to-deliver*. Under physical settlement, the protection buyer may deliver any senior debt obligation of the reference entity. When the triggering credit event is default, all senior obligations of the reference entity generally trade at roughly equal levels, mainly because of the expected recovery rate in a bankruptcy proceeding. However, where the triggering event is restructuring, short-dated bank debt, which has been restructured to give lending banks better pricing and collateral, will trade at a significant premium to longer-dated bonds. The pricing differential between the short-dated, restructured obligations and the longer-dated bonds results in the delivery option held by the protection buyer carrying significant value, as the protection buyer will deliver the cheapest-to-deliver obligation. Under the modified restructuring definition, where the triggering event is restructuring, the delivered obligation cannot have a maturity that is longer than the original maturity date of the credit derivative contract, or more than 30 months after the original maturity date.

[7]Note that a contract may be written in relation to a *reference entity*, which is the corporate or sovereign name, or a *reference obligation*, which is a specific debt obligation of a specific reference entity. Other terms for reference obligation are *reference asset* and *reference credit*. We will use these latter terms interchangeably in the book.

- the settlement mechanism for the contract, whether cash-settled or physically settled;
- (under physical settlement) the deliverable obligation that the protection buyer delivers to the protection seller on the occurrence of a trigger event.

2.2.2 Funded and Unfunded Contracts

Credit derivatives are grouped into *funded* and *unfunded* instruments. In a *funded* credit derivative, typified by a credit-linked note (CLN), the investor in the note is the credit-protection seller and is making an upfront payment to the protection buyer when it buys the note. This upfront payment is the price of the CLN. Thus, the protection buyer is the issuer of the note. If no credit event occurs during the life of the note, the redemption value (par) of the note is paid to the investor on maturity. If a credit event does occur, then on maturity a value less than par will be paid out to the investor. This value will be reduced by the nominal value of the reference asset that the CLN is linked to. The exact process will differ according to whether *cash settlement* or *physical settlement* has been specified for the note. We will consider this later.

In an *unfunded* credit derivative, typified by a credit default swap, the protection seller does not make an upfront payment to the protection buyer. Thus the main difference between funded and unfunded contracts is that in a funded contract, the insurance protection payment is made to the protection buyer at the start of the transaction; if there is no credit event, the payment is returned to the protection seller. In an unfunded contract, the protection payment is made on termination of the contract on occurrence of a triggering credit event. Otherwise it is not made at all. When entering into a funded contract transaction, therefore, the protection seller must find the funds at the start of the trade.

Compared to cash market bonds and loans, unfunded credit derivatives isolate and transfer credit risk. In other words, their value reflects (in theory) only the credit quality of the reference entity. Compare this to a fixed-coupon corporate bond, the value of which is a function of both interest-rate risk and credit quality, and whose return to the investor will depend on the investor's funding costs.[8] The interest-rate risk

---

[8] Funding refers to the cost of funds of the investor. For an AA-rated bank it should be Libid. For a traditional investor such as a pension fund it is more problematic, as the funds are, in theory, invested directly with the pension fund and so acquired 'free'. However, for economic purposes such funds are valued at what rate they could be invested in the money markets. For other investors it will be Libor plus a spread, except for very highly rated market participants, such as the World Bank, which can fund at sub-Libor.

element of the bond can be removed by combining the bond with an interest-rate swap, to create an *asset swap*. An asset swap removes the interest-rate risk of the bond, leaving only the credit quality and the funding aspects of the bond. With an unfunded credit derivative the funding aspect is removed as well, leaving only the credit element. This is because no upfront payment is required, resulting in no funding risk to the protection seller. The protection seller, who is the investor, receives a return that is linked only to the credit quality of the reference entity.

We now consider the individual credit derivative instruments in turn.

2.3 CREDIT DEFAULT SWAPS

We describe first the credit default swap (CDS), the most commonly-traded credit derivative instrument.

2.3.1 Structure

The most common credit derivative is the *credit default swap, credit swap* or *default swap*.[9] This is a bilateral contract that provides protection on the par value of a specified reference asset, with a protection buyer that pays a periodic fixed fee or a one-off premium to a protection seller, in return for which the seller will make a payment on the occurrence of a specified credit event.

The fee is usually quoted as a basis point multiplier of the nominal value, and is generally paid quarterly in arrears. The swap can refer to a single asset, known as the reference asset or underlying asset, a basket of assets, or a reference entity. The default payment can be paid in whatever way suits the protection buyer or both counterparties. For example, it may be linked to the change in price of the reference asset or another specified asset, it may be fixed at a pre-determined recovery rate, or it may be in the form of actual delivery of the reference asset at a specified price. The basic structure is illustrated in Figure 2.1.

The CDS enables one party to transfer its credit risk exposure to another party. Banks may use CDS to trade sovereign and corporate credit spreads without trading the actual assets themselves; for example, someone who has gone long on a CDS (the protection buyer) will gain if the reference asset obligor suffers a rating downgrade or defaults, and can sell the CDS at a profit if he can find a buyer

---

[9]The author prefers the first term, but the other two are observed also.

2. CREDIT DERIVATIVE INSTRUMENTS: PART I

Bank A
Protection buyer

Bank B
Protection seller

Fee or premium →

← Default payment on triggering event

Reference Asset

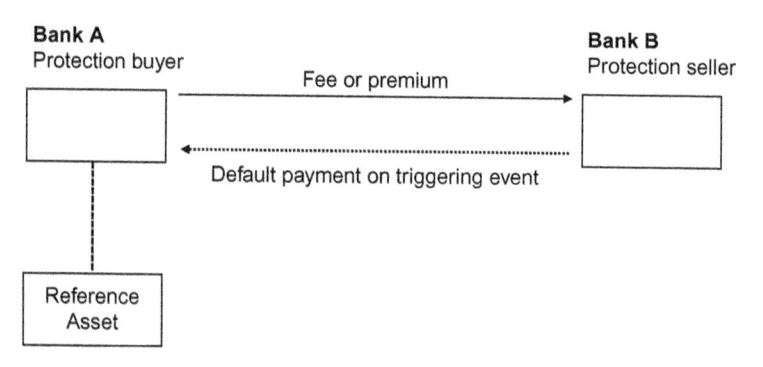

FIGURE 2.1 Credit default swap.

counterparty.[10] This is because the cost of protection on the reference asset will have increased as a result of the credit event. The original buyer of the CDS need never have owned a bond issued by the reference asset obligor.

The maturity of the CDS does not have to match the maturity of the reference asset, and often does not. On occurrence of a credit event, the swap contract is terminated and a settlement payment made by the protection seller or *guarantor* to the protection buyer. This termination value is calculated at the time of the credit event, and the exact procedure that is followed to calculate the termination value will depend on the settlement terms specified in the contract. This will be either cash settlement or physical settlement. We look at these options later.

For illustrative purposes, Figure 2.2(a) shows investment-grade credit default swap levels during 2001 and 2002 for US dollar and Euro reference entities (average levels taken). Figures 2.2(b) and (c) show the picture for 2010–2012 for US and European investment-grade banks (average levels).

A sample CDS term sheet is given in Appendix 2.2.

---

[10] Be careful with the terminology here. To 'go long' on an instrument generally is to purchase it. In the cash market, going long on the bond means one is buying the bond and so receiving coupon; the buyer has therefore taken on credit risk exposure to the issuer. In a CDS, going long is to buy the swap, but the buyer is purchasing protection and therefore paying premium; the buyer has no credit exposure on the name and has in effect 'gone short' on the reference name (the equivalent of shorting a bond in the cash market and paying coupon). So buying a credit default swap is frequently referred to in the market as 'shorting' the reference entity. To avoid confusion, it is always best to speak of either 'buying protection' or 'selling protection'.

2.3 CREDIT DEFAULT SWAPS

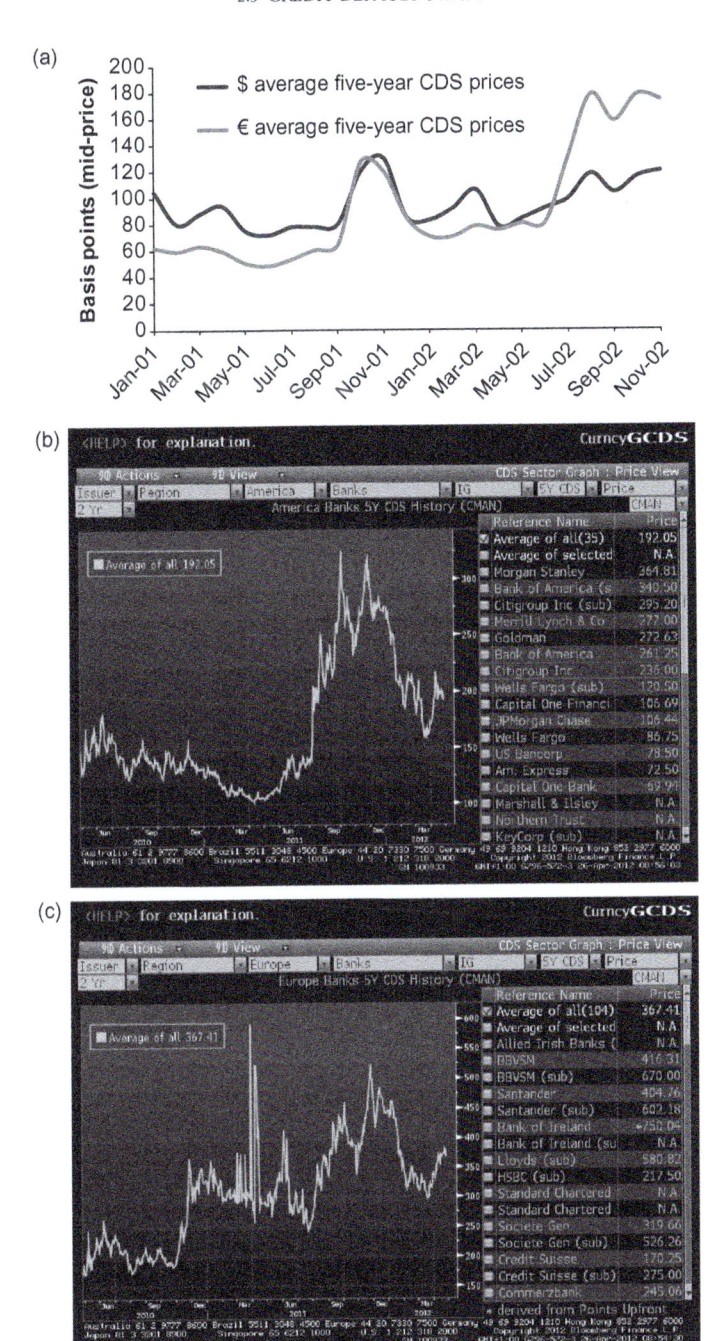

FIGURE 2.2 (a) Investment-grade credit default swap levels, 2001–2002. (b) CDS history, US banks, 2010–2012. (c) CDS history, European banks, 2010–2012. © *Bloomberg LP. Used with permission.*

AN INTRODUCTION TO CREDIT DERIVATIVES

EXAMPLE 2.1: CREDIT DEFAULT SWAP EXAMPLE

XYZ plc credit spreads are currently trading at 120 bps over government for 5-year maturities and 195 bps over for 10-year maturities. A portfolio manager hedges a $10 million holding of 10-year paper by purchasing the following credit default swap, written on the 5-year bond. This hedge protects for the first 5 years of the holding, and in the event of XYZ's credit spread widening will increase in value and may be sold on before expiry at profit. The 10-year bond holding also earns 75 bps over the shorter-term paper for the portfolio manager.

Term 5 years

Reference credit	XYZ plc 5-year bond
Credit event	The business day following occurrence of specified credit event
Default payment	Nominal value of bond × [100 − price of bond after credit event]
Swap premium	3.35%

Assume now that midway into the life of the swap there is a technical default on the XYZ plc 5-year bond, such that its price now stands at $28. Under the terms of the swap the protection buyer delivers the bond to the seller, who pays out $7.2 million to the buyer.

2.3.2 Basket Default Swaps

The simplest CDS is the single-name credit default swap, which references one reference entity or the specific asset of an entity. A basket default swap is linked to a group of reference entities. There may be five, ten, twenty or more reference names in the basket. While it is possible to buy a CDS that covers all the named assets in the event of default, this is rare, and the most common basket CDS provides protection on a selection of the names in the basket only. For instance, if there are q names in the basket, the basket CDS may be one of the following:

- first-to-default, which provides credit protection on the first default in the basket only;
- second-to-default, which provides credit protection on the second default in the basket (but not the first);
- nth-to-default, which provides protection on the first n (out of q) defaults in the basket;
- last pth-to-default, which provides protection on the last p (out of q) defaults.

Basket default swaps are the main building blocks for other types of more complex structured product such as synthetic collateralized debt obligations. These often feature a portfolio of reference assets, all of which are the reference names in one basket CDS.

2.3.3 Unwinding a CDS Position

The CDS should be viewed primarily as an investment instrument. This is subtly different from viewing it as a risk management or risk mitigation tool. The terminology used in the market is a throwback to the days when CDS was considered the same way as an insurance contract. Hence dealing in a CDS is known as buying or selling protection. From an investor viewpoint of course one is either buying a product (to 'go long') or selling a product (to 'go short'). So 'buying a CDS' is often used to refer to buying protection, which is equivalent to going short the asset. Table 2.1 makes the meanings clear.

The best approach to avoid any confusion is to always speak in terms of buying or selling protection, rather than simply buying or selling or going long or short.

A CDS is better viewed as a credit asset in its own right because its value moves in line with credit spreads generally and it is marked-to-market in a bank's trading book. This is unlike an insurance contract and more like a corporate bond floating-rate note (FRN). The premium in a CDS contract, like the spread over Libor in an FRN coupon payment, is fixed at trade inception; however, in the secondary market this premium will change as the credit risk perception of the reference asset changes. In an FRN the coupon spread does not change; rather, the actual price of the bond in the secondary market rises or falls as credit perception improves or worsens.

In the cash bond market there is no issue with 'unwinding' a position. If one is long a bond, one will simply sell it. A short position is

TABLE 2.1 Comparing Cash and Synthetic Markets

CDS Market	CDS Cashflow	Investment Position
Buying protection	Pay (fixed) premium	Short the cash asset
Selling protection	Receive ('floating') premium	Long the cash market
Bond Market	**Bond Cashflow**	**Investment Position**
Buying the bond	Receive fixed or floating coupon	Long the cash market
Selling the bond	Pay fixed or floating coupon	Short the cash asset

flattened out by buying it back. For a CDS, the unwind can be carried out in one of three ways:

- Entering into an offsetting CDS position: a bought/sold protection position is effectively closed out buy entering into a new sold/bought protection CDS position of matching tenor. This crystallizes the mark-to-market profit/loss on the original trade, but note that both trades remain live until maturity on occurrence of a credit event. If the second trade has been entered into with a different counterparty, then the bank will have counterparty exposure in both cases; however, if a credit event does occur, the bank will have counterparty exposure with regard to the bought protection position. This type of offset does not truly unwind the original position.
- Terminate the CDS: the trader will cancel the position with the existing counterparty. The present value of the contract, based on its current mark-to-market value *vis-á-vis* the original trade price, is paid or received by the trader. This closes out the position and the contract, which terminates.
- Novation or assignment: the trader hands the CDS contract to a third-party, which takes over the position of the trader. The trader will pay or receive the current mark-to-market value of the contract, which continues in existence.

Only termination of a contract actually unwinds it; as we observe above, the other two options result in the exposure being retained, albeit in hedged form or residing with another entity.

2.4 ASSET SWAPS

Asset swaps pre-date the introduction of 'credit derivatives' in the market but have subsequently been viewed as part of it. They are really interest-rate swaps (IRS) combined with a cash bond.

2.4.1 Description

An asset swap is a combination of an interest-rate swap and a bond, and is used to alter the cash flow profile of a bond.[11] The asset swap market is an important information point for the credit derivative market since it explicitly sets out the price of credit as a spread over Libor. Pricing a bond by reference to Libor is commonly used and the spread over Libor is a measure of credit risk in the cash flow of the underlying

---

[11]For a background on interest-rate swaps, the reader can look up any number of sources, for instance Das (1994), Kolb (1999), Decovny (1999), Choudhry (2001), and so on.

2.4 ASSET SWAPS

bond. This is because Libor itself — the rate at which banks lend cash to each other in the inter-bank market — is viewed as representing the credit risk of banks. As such it can be viewed as having a generic A-credit rating. The spread over Libor therefore represents additional credit risk over and above that of bank risk. The main reason for entering into an asset swap is to enable the investor to take exposure to the credit quality of a bond with (in theory) zero interest rate risk.

Asset swaps are used to transform the cash flow characteristics of a bond, either fixed-rate into floating-rate or floating-rate into fixed-rate. This enables investors to hedge the currency, credit and interest rate risks to create investments with more suitable cash flow characteristics for themselves. An asset swap package involves transactions in which the investor acquires a bond position and then enters into an interest rate swap with the bank that sold him the bond. If it is a fixed-rate bond, the investor will pay fixed and receive floating on the interest-rate swap. This transforms the fixed coupon of the bond into a Libor-based floating coupon. The generic structure is shown in Figure 2.3.

In an asset swap the asset swap buyer takes on the credit risk of the bond. If the bond defaults, the asset swap buyer has to continue paying on the swap — which can no longer be funded with the coupon from the bond — or the swap can be closed out at market value. The asset swap buyer also loses the par redemption of the bond, receiving whatever recovery rate the bond issuer pays. As a result the buyer has a default contingent exposure to the mark-to-market on the swap and to the redemption on the asset. The buyer is exposed to the loss of the coupons and redemption on the bond — the difference between the bond price and recovery value. In economic terms the purpose of the asset swap spread is to compensate the asset swap buyer for taking these risks.

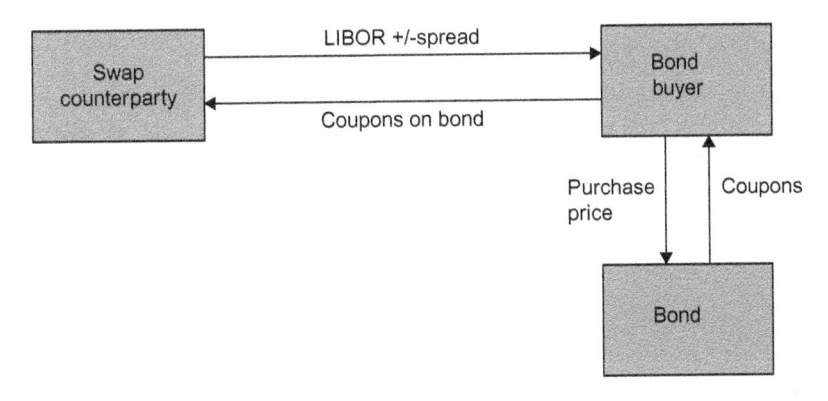

FIGURE 2.3 Asset swap.

2.4.2 Illustration Using Bloomberg

We can illustrate the asset swap spread for a credit-risky corporate bond using Bloomberg screens. In Figure 2.4(a) we show the 7.5% 2016 bond issued by British Telecom plc, a UK telecoms company, in January 2001. The bond is denominated in GBP, and is a form of 'credit-linked note', because its coupon increases by 25 basis points each time the issuer credit rating is downgraded by 1 notch. On issue the bond was rated A-/A3, as at April 2012 it was rated Baa2/BBB. Figure 2.4(a) is Bloomberg's YA page for yield analysis, which is obtained by typing:

BRITEL 7.5 016 < CORP > YA < GO >

and shows the bond as at 26 April 2012, at an offered price of 122.21, which represents a gross redemption yield of 3.22%. Combining this with an interest-rate swap to create an asset swap will convert the bond's fixed coupon to a floating-rate coupon for the bondholder, who pays fixed and receives floating in the associated interest-rate swap.

To see what the return spread for this bond would be in an asset swap, we call up screen ASW. This is shown in Figure 2.4(b) and we see that the asset swap spread for the bond is 181.4 basis points. The bond price on the screen is user-input at 122.21 as before.

2.5 TOTAL RETURN SWAPS

A *total return swap* (TRS), sometimes known as a *total rate of return swap* or *TR swap*, is an agreement between two parties that exchanges the total return from a financial asset between them. This is designed to transfer the credit risk from one party to the other. It is one of the principal instruments used by banks and other financial institutions to manage their credit risk exposure, and as such is a credit derivative. One definition of a TRS is given in Francis *et al.* (1999), which states that a TRS is a swap agreement in which the *total return* of a bank loan or credit-sensitive security is exchanged for some other cash flow, usually tied to Libor or some other loan or credit-sensitive security.

The TRS trade itself can be to any maturity term — that is, it need not match the maturity of the underlying security. In a TRS, the total return from the underlying asset is paid over to the counterparty in return for a fixed or floating cash flow. This makes it slightly different to other credit derivatives, as the payments between counterparties to a TRS are connected to changes in the market value of the underlying asset, as well as changes resulting from the occurrence of a credit event. So, in other words, TRS cash flows are not solely linked to the occurrence of a credit event; in a TRS the interest-rate risk is also transferred. The transaction enables the complete cash flows of a bond to be received without the

2.5 TOTAL RETURN SWAPS 31

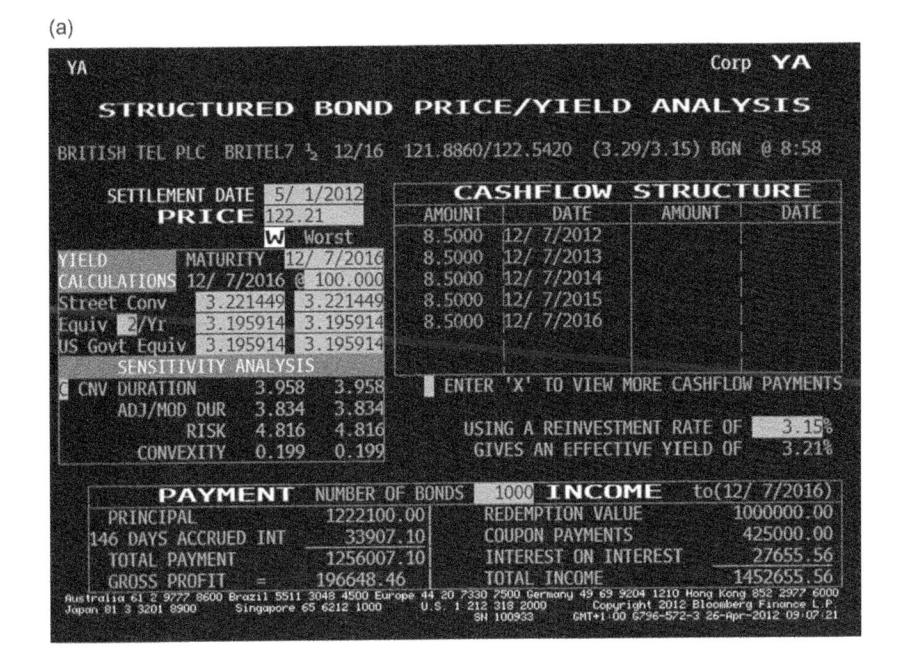

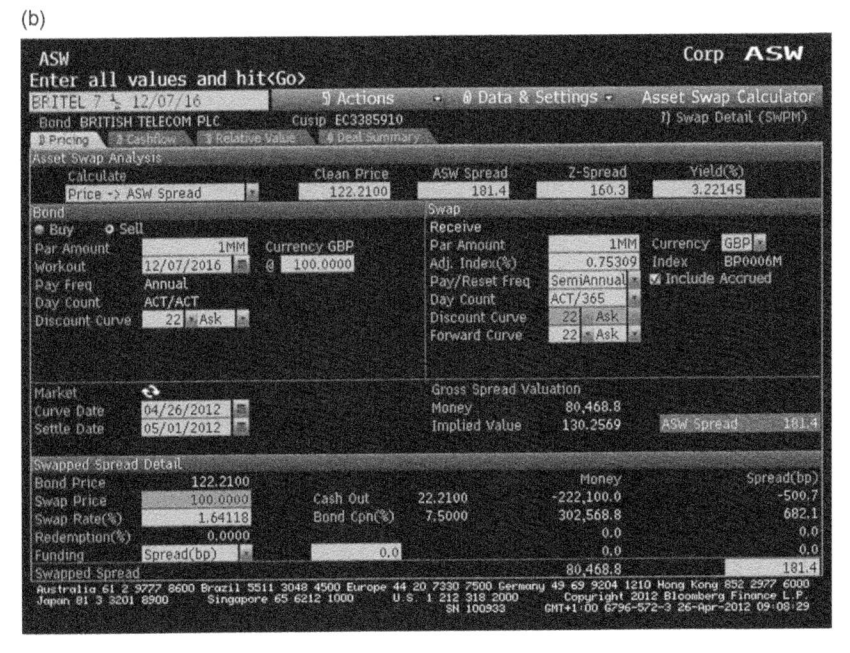

(a)

(b)

FIGURE 2.4 (a) Bloomberg screen YA for British Telecom 7.5% 2016 bond, as at 26 April 2012. (b) Bloomberg screen ASW for British Telecom 7.5% 2016 bond as at April 2012. © *Bloomberg L.P., reproduced with permission. Visit: www.bloomberg.com*

Bank A
Total return payer

Bank B
Total return receiver

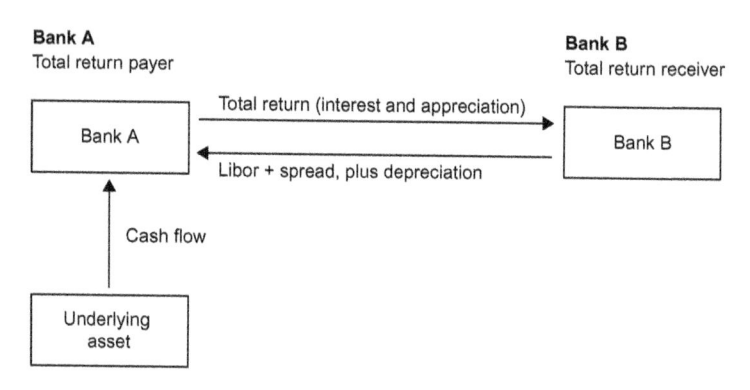

```
                Total return (interest and appreciation)
     Bank A  <------------------------------------------  Bank B
            ------------------------------------------>
                   Libor + spread, plus depreciation

        ^
        | Cash flow
        |
    Underlying
      asset
```

FIGURE 2.5 Total return swap.

recipient actually buying the bond, which makes it a synthetic bond product and therefore a credit derivative. An investor may wish to receive such cash flows synthetically for tax, accounting, regulatory capital, external audit or legal reasons. On the other hand, it may be easier to source the reference asset synthetically — via the TRS — than in the cash market. This happens sometimes with illiquid bonds.

In some versions of a TRS the actual underlying asset is actually sold to the counterparty, with a corresponding swap transaction agreed alongside; in other versions there is no physical change of ownership of the underlying asset. The first would make TRS akin to a synthetic repo transaction. This is discussed in Choudhry (2004).

Figure 2.5 illustrates a generic TR swap. The two counterparties are labelled as banks, but the party termed 'Bank A' can be another financial institution, including cash-rich fixed-income portfolio managers such as insurance companies, and hedge funds. In the figure, Bank A has contracted to pay the 'total return' on a specified reference asset, while simultaneously receiving a Libor-based return from Bank B. The reference or underlying asset can be a bank loan such as a corporate loan or a sovereign or corporate bond. The total return payments from Bank A include the interest payments on the underlying loan, as well as any appreciation in the market value of the asset. Bank B will pay the Libor-based return; it will also pay any difference if there is depreciation in the price of the asset. The economic effect is as if Bank B owned the underlying asset, as such TR swaps are synthetic loans or securities. A significant feature is that Bank A will usually hold the underlying asset on its balance sheet, so that if this asset was originally on Bank B's balance sheet, this is a means by which the latter can have the asset removed from its balance sheet for the term of the TR swap.[12]

---

[12] Although it is common for the receiver of the Libor-based payments to have the reference asset on its balance sheet, this is not always the case.

If we assume Bank A has access to Libor funding, it will receive a spread on this from Bank B. Under the terms of the swap, Bank B will pay the difference between the initial market value and any depreciation, so it is sometimes termed the 'guarantor' while Bank A is the 'beneficiary'.

The total return on the underlying asset is the interest payments and any change in the market value if there is capital appreciation. The value of an appreciation may be cash-settled, or alternatively there may be physical delivery of the reference asset on maturity of the swap, in return for a payment of the initial asset value by the total return 'receiver'. The maturity of the TR swap need not be identical to that of the reference asset, and in fact it is rare for it to be so.

The swap element of the trade will usually pay on a quarterly or semi-annual basis, with the underlying asset being re-valued or *marked-to-market* on the re-fixing dates. The asset price is usually obtained from an independent third-party source, such as Bloomberg or Reuters, or as the average of a range of market quotes. If the *obligor* of the reference asset defaults, the swap may be terminated immediately with a net present value payment changing hands according to what this value is, or it may be continued with each party making appreciation or depreciation payments as appropriate. This second option is only available if there is a market for the asset, which is unlikely in the case of a bank loan. If the swap is terminated, each counterparty will be liable to the other for accrued interest plus any appreciation or depreciation of the asset. Commonly under the terms of the trade, the guarantor bank has the option to purchase the underlying asset from the beneficiary bank, and then deal directly with the loan defaulter.

The TRS can also be traded as a funded credit derivative, and we look at this in Chapter 3.

Banks have employed a number of methods to price credit derivatives and TR swaps. Essentially, the pricing of credit derivatives is linked to that of other instruments; however, the main difference between credit derivatives and other off-balance-sheet products such as equity, currency or bond derivatives is that the latter can be priced and hedged with reference to the underlying asset, which can be problematic when applied to credit derivatives. Credit product pricing uses statistical data on likelihood of default, probability of payout, level of risk tolerance and a pricing model. With a TR swap, the basic concept is that one party 'funds' an underlying asset and transfers the total return of the asset to another party in return for a (usually) floating return that is a spread to Libor. This spread is a function of:

- the credit rating of the swap counterparty;
- the amount and value of the reference asset;
- the credit quality of the reference asset;

- the funding costs of the beneficiary bank;
- any required profit margin;
- the capital charge associated with the TR swap.

The TR swap counterparties must consider a number of risk factors associated with the transaction, which include:

- the TR beneficiary may default while the reference asset has declined in value;
- the possibility that the reference asset obligor defaults, followed by default of the TR swap receiver before payment of the depreciation has been made to the payer or 'provider'.

The first risk measure is a function of the probability of default by the TR swap receiver and the market volatility of the reference asset, while the second risk is related to the joint probability of default of both factors, as well as the recovery probability of the asset.

> ### EXAMPLE 2.2: THE TRS AS A FUNDING INSTRUMENT
>
> TRS contracts are used in a variety of applications by banks, other financial institutions and corporates. As we noted in the main body of the text, they can be written as pure exchanges of cash flow differences - rather like an interest-rate swap - or the reference asset can be actually transferred to the total return payer, which would then make the TRS akin to a synthetic repo contract.[13]
>
> We describe here the use of TRS as a funding instrument, in other words as a substitute for a repo trade.[14] Consider a financial institution such as a regulated broker-dealer that has a portfolio of assets on its balance sheet that it needs to obtain funding for. These assets are investment-grade rated structured finance bonds such as credit card ABS, residential MBS and CDO notes, and investment-grade rated convertible bonds. In the repo

---

[13] The economic effect may be the same, but they are considered different instruments. TRS actually takes the assets off the balance sheet, whereas the tax and accounting authorities treat repo as if the assets remain on the balance sheet. In addition, a TRS trade is conducted under the ISDA standard legal agreement, while repo is conducted under the Global Master Repurchase Agreement (GMRA) standard repo legal agreement. It is these differences that, under certain circumstances, make the TRS funding route a more favourable one.

[14] There may be legal, administrative, operational or other reasons why a repo trade is not entered into to begin with. In these cases, provided that a counterparty can be found and the funding rate is not prohibitive, a TRS may be just as suitable.

2.5 TOTAL RETURN SWAPS

market, it is able to fund these at Libor plus 6 basis points. That is, it can repo the bonds out to a bank counterparty, and will pay Libor plus 6 bps on the funds it receives.

Assume that for operational reasons the bank can no longer fund these assets using repo. It can fund them using a basket TRS instead, providing a suitable counterparty can be found. Under this contract, the portfolio of assets is swapped out to the TRS counterparty, and cash received from the counterparty. The assets are therefore sold off the balance sheet to the counterparty, an investment bank. The investment bank will need to fund these itself, it may have a line of credit from a parent bank or it may swap the bonds out itself. The funding rate it charges the broker-dealer will depend on what rate it can fund the assets itself. Assume this is Libor plus 12 bps — the higher rate reflects the lower liquidity in the basket TRS market for non-vanilla bonds compared to repo. The broker-dealer enters into a 3-month TRS with the investment bank counterparty, with a one-week interest rate reset. This means at each one-week interval, the basket is revalued. The difference in value from the last valuation is paid (if higher) or received (if lower) by the investment bank to the broker-dealer; in return the broker-dealer also pays one-week interest on the funds it received at the start of the trade. In practice these two cash flows are netted off and only one payment changes hands, just like in an interest-rate swap.

The terms of the trade are shown below:

Trade date	22 December 2003
Value date	24 December 2003
Maturity date	24 March 2004
Rate reset	31 December 2003
Interest rate	1.19875% (this is one-week USD Libor fix of 1.07875 plus 12 bps)

The swap is a 3-month TRS with one-week reset, which means that the swap can be broken at one-week intervals and bonds in the reference basket can be returned, added to or substituted. Assume that the portfolio basket contains five bonds, all US dollar denominated.

Assume these are all investment-grade rated credit card ABS bonds with prices available on Bloomberg. The combined market value of the entire portfolio is taken to be USD 151,080,951.00.

At the start of the trade, the five bonds are swapped out to the investment bank, who pays the portfolio value for them. On the first reset date, the portfolio is revalued and the following calculations confirmed:

Old portfolio value USD	151,080,951.00
Interest rate	1.19875%
Interest payable by broker-dealer	USD 35,215.50

New portfolio value	USD 152,156,228.00
Portfolio performance	+1,075,277
Net payment: broker-dealer receives	USD 1,040,061.50

The rate is reset for value 31 December 2003 for the period to 7 January 2004. The rate is 12 bps over the one-week USD Libor fix on 29 December 2003, which is 1.15750 + 0.12 or 1.2775%. This interest rate is payable on the new loan amount of USD 152,156,228.00.

2.6 INDEX CDS: THE ITRAXX INDEX

The iTraxx series is a set of credit indices that enable market participants to trade funded and unfunded credit derivatives linked to a credit benchmark. There are a number of different indices covering different sectors, for example iTraxx Europe, iTraxx Japan, iTraxx Korea, and so on. The equivalent index in the North American market is known as CD-X. The iTraxx exhibits relatively high liquidity and for this reason is viewed as a credit benchmark, and its bid-offer spread is very narrow at 1–2 basis points. This contrasts with spreads generally between 10 and 30 basis points for single-name CDS contracts. Because of its liquidity and benchmark status, the iTraxx is increasingly viewed as a leading indicator of the credit market overall, and the CDS index basis is important in this regard as an indicator of relative value.

The iTraxx series is a basket of reference credits that is reviewed on a regular basis. For example, the iTraxx Europe index consists of 125 corporate reference names, so that each name represents 0.8% of the basket. Figure 2.6 shows the an extract from a Bloomberg screen for the June 2011 iTraxx Europe index, with the first page of reference names. Figure 2.7 shows additional terms for this index contract.[15]

The index rolls every 6 months (in March and September), when reference names are reviewed and the premium is set. Hence there is a rolling series of contracts with the 'front contract' being the most recent. There are two standard maturities, which are 5.25 years and 10.25 years. Figure 2.8 shows a list of iTraxx indices as at June 2006; the second-listed contract is the current one, with a June 2011 maturity and a premium of 40 basis points (see Figure 2.6). All existing indices can be traded although the most liquid index is the current one. Reference names are all investment-grade rated and are the highest traded names by CDS volume in the past 6 months.

A bank buying protection in EUR 10 million notional of the index has in effect bought protection on EUR 80,000 each of 125 single-name

---

[15]The screens for the iTraxx are found by typing ITRX CDS <Corp> <go>.

2.6 INDEX CDS: THE ITRAXX INDEX 37

```
<HELP> for explanation, <MENU> for similar functions.        P174 Corp

      CREDIT  DEFAULT  SWAPS     for ticker ITRX  CDS Page   1/ 11
                                                       Found   194

      ISSUER          SPREAD  MATURITY  SERS RTNG  FREQ   TYPE    CNTRY/CURR
  1)  ITRX EUR          25    6/20/09   5EU  N.A.  Qtr   iTRAXX    EU  /EUR
  2)  ITRX EUR          40    6/20/11   5EU2 N.A.  Qtr   iTRAXX    EU  /EUR
  3)  ITRX EUR          50    6/20/13   5EU3 N.A.  Qtr   iTRAXX    EU  /EUR
  4)  ITRX EUR          60    6/20/16   5EU4 N.A.  Qtr   iTRAXX    EU  /EUR
  5)  ITRX SDI          45    6/20/16   2SD  N.A.  Qtr   iTRAXX    GB  /GBP
  6)  ITRX INDS         40    6/20/11   5IND N.A.  Qtr   iTRAXX    EU  /EUR
  7)  ITRX INDS         60    6/20/16   5IN2 N.A.  Qtr   iTRAXX    EU  /EUR
  8)  ITRX SUB          25    6/20/11   5SUB N.A.  Qtr   iTRAXX    EU  /EUR
  9)  ITRX SUB          45    6/20/16   5SU2 N.A.  Qtr   iTRAXX    EU  /EUR
 10)  ITRX SNR FIN      15    6/20/11   5SNR N.A.  Qtr   iTRAXX    EU  /EUR
 11)  ITRX SNR FIN      25    6/20/16   5SN2 N.A.  Qtr   iTRAXX    EU  /EUR
 12)  ITRX CROSS       290    6/20/11   5XOV N.A.  Qtr   iTRAXX    EU  /EUR
 13)  ITRX CROSS       350    6/20/16   5XO2 N.A.  Qtr   iTRAXX    EU  /EUR
 14)  ITRX NON-FINL     40    6/20/11   5NF1 N.A.  Qtr   iTRAXX    EU  /EUR
 15)  ITRX NON-FINL     60    6/20/16   5NF2 N.A.  Qtr   iTRAXX    EU  /EUR
 16)  ITRX TMT          40    6/20/11   5TMT N.A.  Qtr   iTRAXX    EU  /EUR
 17)  ITRX TMT          60    6/20/16   5TM2 N.A.  Qtr   iTRAXX    EU  /EUR
 18)  ITRX HVOL         40    6/20/09   5HI  N.A.  Qtr   iTRAXX    EU  /EUR
 19)  ITRX HVOL         70    6/20/11   5HI2 N.A.  Qtr   iTRAXX    EU  /EUR
Australia 61 2 9777 8600     Brazil 5511 3048 4500      Europe 44 20 7330 7500    Germany 49 69 920410
Hong Kong 852 2977 6000 Japan 81 3 3201 8900 Singapore 65 6212 1000 U.S. 1 212 318 2000 Copyright 2006 Bloomberg L.P.
                                                                                    0 19-Jun-06 15:35:31
```

FIGURE 2.6 List of iTraxx indices as shown on Bloomberg, 19 June 2006.

```
<HELP> for explanation.                                    P174 Corp   CDSW
<Menu> to return

      ADDITIONAL  DESCRIPTIVE  INFORMATION

         Announcement Date:  0/ 0/00      Currency:       EUR
         Int. Accrual Date:  3/20/06
                                          Amt Issued:      0.00
            1st Settle Date: 3/20/06   Amt Outstanding:    0.00
            1st Coupon Date: 6/20/06
                                          Par Amount:      0.00
              Maturity Date: 6/20/11

         Payment Frequency: Q Quarterly
           Day Count Basis: ACT/360

              Business Days: EUR
           Business Day Adj: 1 Following

              Issue Spread:   40.0 bps


Australia 61 2 9777 8600     Brazil 5511 3048 4500      Europe 44 20 7330 7500    Germany 49 69 920410
Hong Kong 852 2977 6000 Japan 81 3 3201 8900 Singapore 65 6212 1000 U.S. 1 212 318 2000 Copyright 2006 Bloomberg L.P.
                                                                                    0 19-Jun-06 15:38:55
```

FIGURE 2.7 Additional terms for June 2011 iTraxx Europe index.

```
<HELP> for explanation.                                    P174 Corp   CDSW
1<GO> to sort by name. 3<Go> to sort by weight. 4<Go> to download to Excel.
                                                              Page 1/7
┌──────────────── REFERENCE ENTITY LIST ─────────────────────────────┐
│ Reference Entity Legal Name              Weight (%)                │
│ ABN AMRO Bank N.V.                         0.800                   │
│ ACCOR                                      0.800                   │
│ Adecco S.A.                                0.800                   │
│ Aegon N.V.                                 0.800                   │
│ Aktiebolaget Electrolux                    0.800                   │
│ Aktiebolaget Volvo                         0.800                   │
│ AKZO Nobel N.V.                            0.800                   │
│ Allianz Aktiengesellschaft                 0.800                   │
│ ALTADIS, S.A.                              0.800                   │
│ ARCELOR FINANCE                            0.800                   │
│ ASSICURAZIONI GENERALI - SOCIETA PER A     0.800                   │
│ AVIVA PLC                                  0.800                   │
│ AXA                                        0.800                   │
│ BAA PLC                                    0.800                   │
│ BAE SYSTEMS PLC                            0.800                   │
│ BANCA INTESA S.P.A.                        0.800                   │
│ BANCA MONTE DEI PASCHI DI SIENA S.P.A.     0.800                   │
│ BANCA POPOLARE ITALIANA - BANCA POPOLA     0.800                   │
│ BANCO BILBAO VIZCAYA ARGENTARIA, SOCIE     0.800                   │
│ Banco Comercial Portugues, S.A.            0.800                   │
Australia 61 2 9777 8600     Brazil 5511 3048 4500    Europe 44 20 7330 7500    Germany 49 69 920410
Hong Kong 852 2977 6000  Japan 81 3 3201 8900  Singapore 65 6212 1000  U.S. 1 212 318 2000  Copyright 2006 Bloomberg L.P.
                                                                    O 19-Jun-06 15:39:20
```

FIGURE 2.8 Page 1 of list of reference names in iTraxx Europe June 2011 index.

CDS. The premium payable on a CDS written on the index is set at the start of the contract and remains fixed for its entire term; the premium is paid quarterly in arrears in the same way as a single-name CDS. The premium remains fixed but of course the market value fluctuates on a daily basis. This works as follows:

- The constituents of the index are set about one week before it goes live, with the fixed premium being set 2 days before. The premium is calculated as an average of all the premiums payable on the reference names making up the index. In June 2006 the current 5-year index for Europe was the iTraxx Europe June 2011 contract. The reference names in the index were set on 13 March 2006, with the premium fixed on 18 March 2006. The index went live on 20 March 2006. The index is renewed every 6 months in the same way.
- After the roll date, a trade in the iTraxx is entered into at the current market price.
- Because this is different to the fixed premium, an up-front payment is made between the protection seller and protection buyer, which is the difference between the present values of the fixed premium and the current market premium.

So for example, on 21 June 2006 the market price of the June 2011 iTraxx Europe was 34 basis points. An investor selling protection on this contract would receive 40 basis points quarterly in arrears for the

2.6 INDEX CDS: THE ITRAXX INDEX

```
<HELP> for explanation.                              P174 Corp    CDSW
2<GO> to save curve source
                   CREDIT DEFAULT SWAP                         CPU:121
```

Deal Information		Spreads	Term
Reference: ITRAXX EUROPE		Curve Date: 6/19/06	
Counterparty: ITRX EUR	Deal#: SPN5ZTGM	Benchmark: S 45 Ask	
Ticker: ITRX CDS Series: 5EU2		EU BGN Swap Curve	
Business Days: EUR	Settlement Code: EUR	Sprds: U User	Ask
Business Day Adj: 1 Following	Currency: EUR	CDSD SPN5ZTGM	IMM N
B BUY Notional: 10.00 MM	Factor:1		
Effective Date: 3/20/06	Knock Out: N	Par Cds Spreads	Default
Maturity Date: 6/20/11	Day Count: ACT/360	Flat: Y (bps)	Prob
Payment Freq: Q Quarterly	Month End: N	6 mo 34.000	0.0029
Pay Accrued: T True	First Cpn: 6/20/06	1 yr 34.000	0.0057
Curve Recovery: T True Next to Last Cpn: 3/21/11		2 yr 34.000	0.0114
Recovery Rate: 0.40 Date Gen Method: B Backward		3 yr 34.000	0.0171
Deal Spread: 40.000 bps		4 yr 34.000	0.0226
Calculator	Mode: C Calc Price	5 yr 34.000	0.0282
Settlement Date: 6/20/06	Model: J JPMorgan	7 yr 34.000	0.0393
Cash Settled On: 6/22/06		10 yr 34.000	0.0556
Price: 100.27280594 Repl Sprd:34.000 bps		Frequency: Q Quarterly	
Market Val: -27,280.59 Days: 0		Day Count: ACT/360	
Accrued: 0.00 Sprd DV01:4,557.82		Recovery Rate: 0.40	
Total Val: -27,280.59 IR DV01: 6.71			

```
Australia 61 2 9777 8600    Brazil 5511 3048 4500    Europe 44 20 7330 7500    Germany 49 69 920410
Hong Kong 852 2977 6000  Japan 81 3 3201 8900  Singapore 65 6212 1000  U.S. 1 212 318 2000  Copyright 2006 Bloomberg L.P.
                                                                              0 19-Jun-06 15:37:06
```

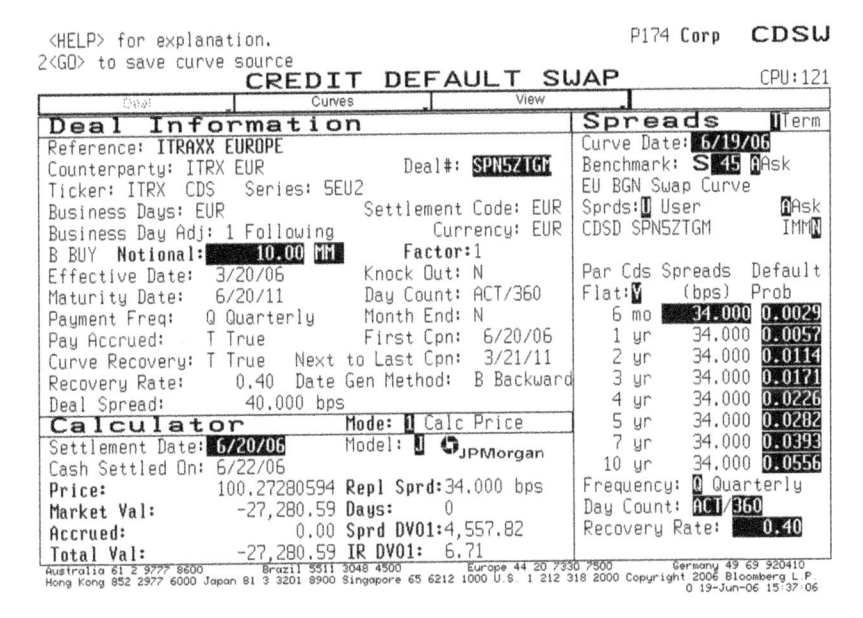

FIGURE 2.9 Screen CDSW used to calculate up-front present value payment for trade in EUR 10 million notional iTraxx Europe index CDS contract, 19 June 2006.

5 years from June 2006 to June 2011. The difference is made up front: the investor receives 40 basis points although the market level at time of trade is 34 basis points. Therefore, the protection seller makes a one-off payment of the difference between the two values, discounted. The present value of the contract is calculated assuming a flat spread curve and a 40% recovery rate. We can use Bloomberg screen CDSW to work this out, and Figure 2.9 shows such a calculation using this screen. This shows a trade for EUR 10 million notional of the current iTraxx Europe index on 19 June 2006. We see the deal spread is 40 basis points; we enter the current market price of 34 basis points, and assume a flat credit term structure.

From Figure 2.9 we see that the one-off payment for this deal is EUR 27,280. The protection seller, who will receive 40 basis points quarterly in arrears for the life of the deal, pays this amount at trade inception to the protection buyer.[16]

If a credit event occurs on one of the reference entities in the iTraxx, the contract is physically settled, for that name, for 0.8% of the notional

---

[16] The one-off payment reflects the difference between the prevailing market rate and the fixed rate. If the market rate was above 40 basis points at the time of this trade, the protection buyer would pay the protection seller the one-off payment reflecting this difference.

value of the contract. This is similar to the way that a single-name CDS would be settled. Unlike a single-name CDS, the contract continues to maturity at a reduced notional amount. Note that European iTraxx indices trade under modified-modified restructuring (MMR) terms, which is prevalent in the European market. Under MMR, a debt restructuring is named as a credit event.[17]

2.7 SETTLEMENT

Credit derivative settlement can follow one of two routes, specified at deal inception. We consider these here.

With all credit derivatives, upon occurrence of a credit event, a credit event notice must be submitted. Typically, the notice must be supported by information posted on public news systems such as Bloomberg or Reuters. When used as part of a structured product, the terms of the deal may state that a credit event must be verified by a third-party *verification agent*. Upon verification, the contract will be settled in one of two ways: cash settlement or physical settlement.

A report from the British Bankers' Association (BBA) suggested that between 75% and 85% of credit derivatives written in 2002 were physically settled, while about 10%–20% were cash settled. In 2010, this position had more or less reversed. About 5% of contracts were settled under the *fixed amount* approach, under which the protection seller delivers a pre-specified amount to the protection buyer ahead of the determination of the reference asset's recovery value. However, as the fixed amount approach is essentially cash settlement, we will consider it as such and prefer the more technical term for it noted below.

2.7.1 Contract Settlement Options

Credit derivatives have a given maturity, but will terminate early if a credit event occurs. On the occurrence of a credit event, the swap contract is terminated and a settlement payment is made by the protection seller or guarantor to the protection buyer. This termination value is calculated at the time of the credit event, and the procedure that is followed to calculate the termination value will depend on the settlement terms specified in the contract. Credit derivatives specify physical or cash settlement. In physical settlement, the protection buyer transfers to the protection seller the deliverable obligation (usually the reference asset or assets), with the total principal outstanding equal to the

---

[17]This contrasts with the North American market, which includes the CDX family of indices, where CDSs trade under no-restructuring terms; this describes only bankruptcy and liquidation as credit events.

2.7 SETTLEMENT

Cash settlement

Protection buyer ← 100% minus recovery rate — Protection seller

Physical settlement

Protection buyer — Reference asset → Protection seller
              ← 100% minus recovery rate —

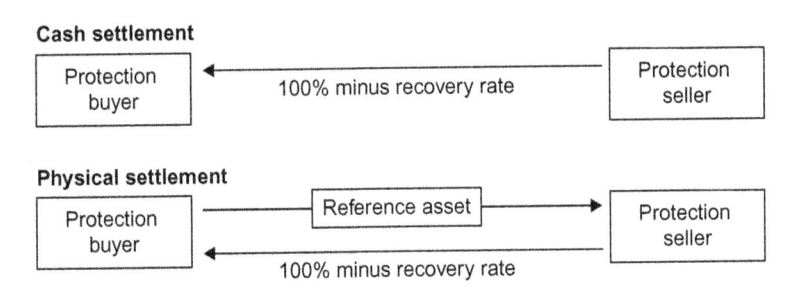

FIGURE 2.10 Cash and physical settlement.

nominal specified in the default swap contract. The protection seller simultaneously pays to the buyer 100% of the nominal. In cash settlement, the protection seller hands to the buyer the difference between the nominal amount of the default swap and the final value for the same nominal amount of the reference asset. This final value is usually determined by means of a poll of dealer banks. This final value is in theory the recovery value of the asset; however as the recovery process can take some time, often the reference asset market value at time of default is taken and this amount used in calculating the final settlement amount paid to the protection buyer.

The settlement mechanisms are shown in Figure 2.10 and follow the following process:

- *Cash settlement*: the contract may specify a pre-determined payout value on occurrence of a credit event. This may be the nominal value of the swap contract. Such a swap is known as a *fixed amount* contract or, in some markets, as a *digital credit derivative*. Alternatively, the termination payment is calculated as the difference between the nominal value of the reference asset and either its market value at the time of the credit event or its recovery value. This arrangement is more common with cash-settled contracts.[18]
- *Physical settlement*: on occurrence of a credit event, the buyer delivers the reference asset to the seller, in return for which the seller pays the face value of the delivered asset to the buyer. The contract may specify a number of alternative assets that the buyer can deliver; these are known as *deliverable obligations*. This may apply when a swap has been entered into on a reference name rather than a

---

[18] Determining the market value of the reference asset at the time of the credit event may be a little problematic: the issuer of the asset may well be in default or administration. An independent third-party *calculation agent* is usually employed to make the termination payment calculation.

specific obligation (such as a particular bond) issued by that name. Where more than one deliverable obligation is specified, the protection buyer will invariably deliver the asset that is the cheapest on the list of eligible assets. This gives rise to the concept of the *cheapest-to-deliver*, as encountered with government bond futures contracts, and is in effect an embedded option afforded the protection buyer.

In theory, the value of protection is identical irrespective of which settlement option is selected. However, under physical settlement the protection seller can gain if there is a recovery value that can be extracted from the defaulted asset, or its value may rise as the fortunes of the issuer improve.

Swap market-making banks often prefer cash settlement as there is less administration associated with it, since there is no delivery of a physical asset. For a CDS used as part of a structured product, cash settlement may be more suitable because such vehicles may not be set up to take delivery of physical assets. Another advantage of cash settlement is that it does not expose the protection buyer to any risks should there not be any deliverable assets in the market, for instance due to shortage of liquidity in the market — were this to happen, the buyer might find the value of its settlement payment reduced. Nevertheless, physical settlement is widely used because counterparties wish to avoid the difficulties associated with determining the market value of the reference asset under cash settlement.[19] Physical settlement also permits the protection seller to take part in the creditor negotiations with the reference entity's administrators, which may result in improved terms for them as holders of the asset.

Cash settlement is sometimes proceeded with even for physically settled contracts when, for one reason or another, it is not possible to deliver a physical asset, for instance if none is available.

2.7.2 Market Requirements

Various market participants have different requirements, and so may have their own preferences with regard to the settlement mechanism. A protection seller may prefer physical settlement for

---

[19] Credit derivative market makers may value two instruments written on the same reference entity, and with all other terms and conditions identical except that one is cash settled and the other physically settled, at the same price. This is because while the protection buyer has a delivery option and will deliver the cheapest bond available, an option that carries value, in a cash-settled contract the protection buyer will nominate this same bond to be used in the calculation of the settlement of the contract. So the value of the delivery option may not result in a higher price quote from a market maker for a physically delivered contract.

2.7 SETTLEMENT

particular reference assets if it believes that a higher recovery value for the asset can be gained by holding onto it and/or entering into the administration process. A protection buyer may have different interests. For instance, unless the protection buyer already holds the deliverable asset (in which case the transaction he has entered into is a classic hedge for an asset already owned), he may prefer cash settlement if he has a negative view of the reference obligation and has used the CDS or other credit derivative to create a synthetic short bond position. Or the protection buyer may prefer physical settlement because he views the delivery option as carrying some value.

2.7.3 Cash Settlement Mechanics

Cash settlement requires a Credit Event Notice and if specified in the related confirmation, a Notice of Publicly Available Information. Generally the cash settlement amount is calculated using market prices for defaulted reference obligations, set by a dealer poll or 'auction' amongst CDS market makers. The seller pays the buyer the notional amount of the trade (floating rate calculation amount) multiplied by the loss of value of the defaulted reference obligations. By doing so, the seller covers the loss of value of the reference obligation caused by the credit event.

More formally, the cash settlement amount is the floating rate payer calculation amount multiplied by the reference price minus the final price, where the floating rate payer calculation amount is the notional amount of the transaction and the final price is the price of the reference obligation. The final price is determined through a valuation method and the parties choose, at the time of the trade, between methods based on 'market value' or 'highest quotations'.

The most common method used for setting the settlement value of the defaulted reference obligation is via a dealer poll of five dealers, whereby:

- The valuation date is agreed at the time of executing the contract, but could be up to 122 days after the credit event.
- It is also possible to use multiple valuation dates.
- The final price is determined by the highest bid price for a specified notional of bonds, and this price is used to determine the compensation amount.
- This final amount is **paid 5 days** after the dealer poll.

The seller of protection is not left with a residual exposure to the defaulted entity.

2.8 RISKS IN CREDIT DEFAULT SWAPS

To conclude this chapter, we consider some risk exposures that investors take on when trading in credit derivatives.

2.8.1 Unintended Risks in Credit Default Swaps

As credit derivatives can be tailored to specific requirements in terms of reference exposure, term to maturity, currency and cash flows, they have enabled market participants to establish exposure to specific entities without the need for them to hold the bond or loan of that entity. This has raised issues of the different risk exposure that this entails compared to the cash equivalent. A Moody's special report highlights the unintended risks of holding credit exposures in the form of default swaps and credit-linked notes (Tolk, 2001). Under certain circumstances it is possible for credit default swaps to create unintended risk exposure for holders, by exposing them to greater frequency and magnitude of losses compared to that suffered by a holder of the underlying reference credit.

In a credit default swap, the payout to a buyer of protection is determined by the occurrence of credit events. The definition of a credit event sets the level of credit risk exposure of the protection seller. A wide definition of 'credit event' results in a higher level of risk. To reduce the likelihood of disputes, counterparties can adopt the ISDA credit derivatives definitions to govern their dealings. The Moody's paper states that the current ISDA definitions do not unequivocally separate and isolate credit risk, and in certain circumstances credit derivatives can expose holders to additional risks. A reading of the paper would appear to suggest that differences in definitions can lead to unintended risks being taken on by protection sellers. Two examples from the paper are cited below as illustration.

2.8.2 Extending Loan Maturity

The bank debt of Conseco, a corporate entity, was restructured in August 2000. The restructuring provisions included deferment of the loan maturity by 3 months, higher coupon, corporate guarantee and additional covenants. Under the Moody's definition, as lenders received compensation in return for an extension of the debt, the restructuring was not considered to be a 'diminished financial obligation', although Conseco's credit rating was downgraded one notch. However, under the ISDA definition the extension of the loan maturity meant that the restructuring was considered to be a credit event, and thus triggered payments on default swaps written on Conseco's bank debt. Hence this was an example of a loss event under ISDA definitions that was not considered by Moody's to be a default.

It was the Conseco case that led to the adoption of the modified restructuring ISDA definitions of 2003.

2.8.3 Risks of Synthetic Positions and Cash Positions Compared

Consider two investors in XYZ, one of whom owns bonds issued by XYZ while the other holds CLN referenced to XYZ. Following a deterioration in its debt situation, XYZ violates a number of covenants on its bank loans, but its bonds are unaffected. XYZ's bank accelerates the bank loan, but the bonds continue to trade at 85 cents on the dollar, coupons are paid and the bond is redeemed in full at maturity. However, the default swap underlying the CLN cites 'obligation acceleration' (of either bond or loan) as a credit event, so the holder of the CLN receives 85% of par in cash settlement and the CLN is terminated. However, the cash investor receives all the coupons and the par value of the bonds on maturity.

These two examples illustrate how, as credit default swaps are defined to pay out in the event of a very broad range of definitions of a 'credit event', portfolio managers may suffer losses as a result of occurrences that are not captured by one or more of the ratings agencies' rating of the reference asset. This results in a potentially greater risk for the portfolio manager compared to the position were it actually to hold the underlying reference asset. Essentially, therefore, it is important for the range of definitions of a 'credit event' to be fully understood by counterparties, so that holders of default swaps are not taking on greater risk than is intended.

2.9 IMPACT OF THE 2007–2008 FINANCIAL CRASH: NEW CDS CONTRACTS AND THE CDS 'BIG BANG'

One of the impacts of the 2007–2008 financial market crisis was that CDS prices rose to hitherto unseen astronomically high levels. The bankruptcy of Lehman Brothers also highlighted the issue of counterparty risk for those market participants that had bought protection using CDS.

One response to this was that the markets changed the protocol for quoting CDS contracts traded in the USA and Canada ('Big Bang') and Europe and Asia-Pacific ('Small Bang').

2.9.1 The CDS 'Big Bang'

ISDA introduced a new supplement and protocol (the 'Big Bang' protocol) and a new standard North American corporate CDS contract with effect from 8 April 2009, the Standard North American Contract (SNAC). The ISDA supplement applied to new CDS transactions. It established credit determination committees, added auction settlement provisions and created backstop dates for credit and succession events. The Big Bang protocol applies to existing CDS transactions. The ISDA SNAC, also referred to as 100/500, applies to North American names denominated in any currency. These CDS contracts trade with an upfront payment and fixed coupons of either 100 basis points for investment-grade reference names and 500 basis points for lower-rated names. The new contract is referred to as the 'SNAC' or the '100/500' contract.

Investment-grade CDS traded with a 100 bps premium are quoted using a flat credit curve. The high-yield names trading at a 500 bps premium are quoted with points upfront. All trades now have a full first coupon with no long or short stub periods. The first accrual start date no longer coincides with the effective date. The effective date was changed to reflect the 'look back' period of 60 days for credit events.[20] The look back periods ensure that offsetting transactions have the same terms and allows positions to be fully hedged.

The ISDA Big Bang protocol applies to existing transactions. It enables market participants to amend outstanding trades so that they can eliminate distinctions between trades entered into before and after 8 April 2009.

2.9.2 CDS and Points Upfront

The 2007–2008 credit crunch resulted in the CDS price for many reference named trading at very high levels; for example Morgan Stanley traded at over 1300 bps and AIG at 1942 bps in September 2008.

---

[20] The look back period is 90 days for succession credit events.

2.9 IMPACT OF THE 2007–2008 FINANCIAL CRASH

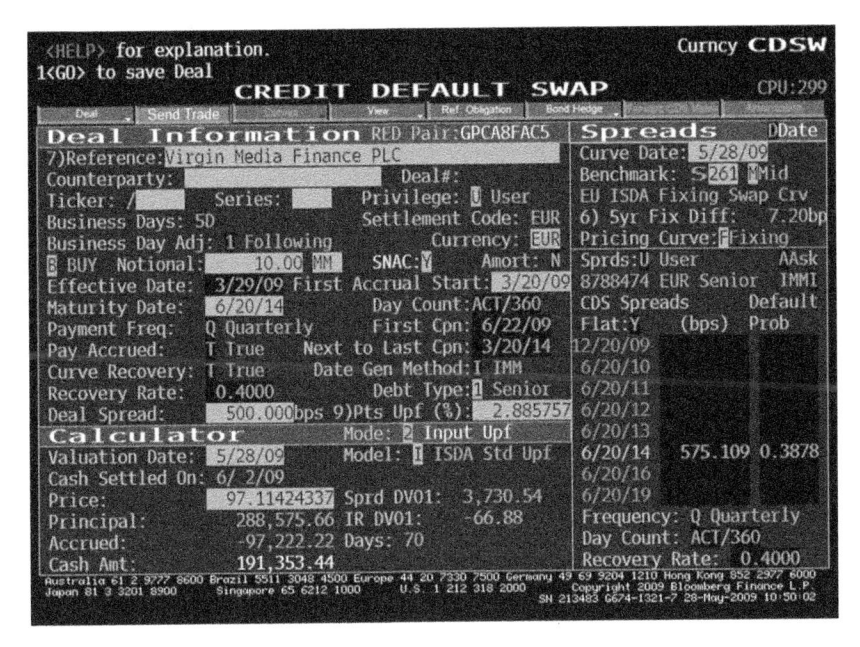

FIGURE 2.11 Bloomberg screen CDSW showing 5-year CDS written on Virgin Media Finance plc name, traded 28 May 2009, with 'points upfront' valuation of 2.88%.

For high risk reference names, CDS spreads that have widened to a large extent are quoted by market makers with 'points upfront'. In this case, if a CDS trades with an upfront fee, a market counterparty buying protection must make an initial payment (a percentage of the notional contract value) as well as a running spread of 500 basis points.

The Bloomberg screen CDSW can be used to value CDS that are quoted with upfront fees, as shown in Figure 2.11. The trade example here is a 5-year CDS quoted on 28 May 2009 on Virgin Media Finance plc. The CDS premium is 575 basis points, and if this traded in the US market there is now a 'points upfront' fixed fee to be paid on inception. This is shown on the screen as 2.885757%. Note also that the pricing model selected has been changed from the hitherto-standard JPMorgan model to the 'ISDA standard upfront' model. Note that in the field 'SNAC' the user has selected 'Y' for yes, indicating this contract is being traded in the US market and not in Europe.

Another response to the market that can be observed from Bloomberg screen CDSW concerns the recovery rate parameter. Previously this had defaulted to 40%. For a large number of lower-rated names this value has been tailor-set to levels ranging from 5% upwards.

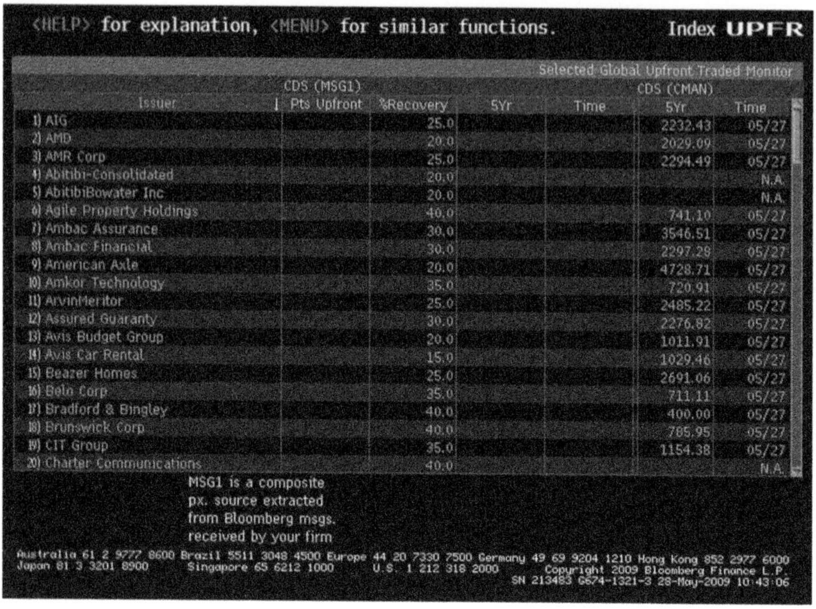
<HELP> for explanation, <MENU> for similar functions. Index UPFR

Selected Global Upfront Traded Monitor

Issuer	CDS (MSG1) Pts Upfront	%Recovery	5Yr	Time	CDS (CMAN) 5Yr	Time
1) AIG		25.0			2232.43	05/27
2) AMD		20.0			2029.09	05/27
3) AMR Corp		25.0			2294.49	05/27
4) Abitibi-Consolidated		20.0				N.A.
5) AbitibiBowater Inc		20.0				N.A.
6) Agile Property Holdings		40.0			741.10	05/27
7) Ambac Assurance		30.0			3546.51	05/27
8) Ambac Financial		30.0			2297.29	05/27
9) American Axle		20.0			4728.71	05/27
10) Amkor Technology		35.0			720.91	05/27
11) ArvinMeritor		25.0			2485.22	05/27
12) Assured Guaranty		30.0			2276.82	05/27
13) Avis Budget Group		20.0			1011.91	05/27
14) Avis Car Rental		15.0			1029.46	05/27
15) Beazer Homes		25.0			2691.06	05/27
16) Belo Corp		35.0			711.11	05/27
17) Bradford & Bingley		40.0			400.00	05/27
18) Brunswick Corp		40.0			785.95	05/27
19) CIT Group		35.0			1154.38	05/27
20) Charter Communications		40.0				N.A.

MSG1 is a composite px. source extracted from Bloomberg msgs. received by your firm

Australia 61 2 9777 8600 Brazil 5511 3048 4500 Europe 44 20 7330 7500 Germany 49 69 9204 1210 Hong Kong 852 2977 6000
Japan 81 3 3201 8900 Singapore 65 6212 1000 U.S. 1 212 318 2000 Copyright 2009 Bloomberg Finance L.P.
SN 213483 G674-1321-3 28-May-2009 10:43:06

FIGURE 2.12 Bloomberg screen UPFR showing reference name recovery rates.

Screen UPFR on Bloomberg shows the recovery rate for selected reference names, now that the market no longer defaults automatically to 40%. Page 1 of this screen is shown at Figure 2.12. We see that recovery rates for this group of companies range from 15% to 40%.

2.9.3 Contract Changes

We summarize below the changes made in CDS trading convention after the events of 2007–2008.

Contract structure:	Auction settlement
	60-day look-back period for credit events
	Maintain quarterly rolls (March, June, September, December)
	Points upfront
	Standard coupons (100/500), with no requirement to reset coupons at each quarterly roll date as both 100 bps and 500 bps are quoted for the life of the contract
Credit events:	Restructuring no longer a credit event
	Note that emerging market (EM) CDS contracts retain restructuring as a credit event

References

Bomfin, A. (2003). *Understanding credit derivatives*. San Diego: Elsevier Academic Press.
Choudhry, M. (2001). *Bond market securities*. London: FT Prentice Hall.
Choudhry, M. (2004). *Structured credit products: credit derivatives and synthetic securitisation*. Singapore: John Wiley & Sons (Asia).
Das, S. (1994). *Swaps and financial derivatives*. London: IFR Publishing.
Das, S. (Ed.), (1998). *Credit derivatives: products, applications and pricing* Singapore: John Wiley & Sons, (Asia).
Das, S. (Ed.), (2000). *Credit derivatives and credit linked notes* Singapore: John Wiley & Sons, (Asia).
Decovny, S. (1999). *Swaps*. Upper Saddle River, NJ: FT Prentice Hall.
Francis, J., Frost, J., & Whittaker, J. (1999). *The handbook of credit derivatives*. New York: McGrawHill.
Gregory, J. (ed.) *Credit derivatives: the definitive guide*. RISK Publishing, London..
Kolb, R. (1999). *Futures, options and swaps*. Oxford: Blackwell.
Schonbucher, P. (2003). *Credit derivatives pricing models: model, pricing and implementation*. Chichester: Wiley Finance.
Tolk, J. (2001) Understanding the risks in credit default swaps, Moody's investors' service special report, 16 March 2001.

APPENDICES

APPENDIX 2.1 ISDA 2002 CREDIT DERIVATIVE DEFINITIONS

Bankruptcy A reference entity voluntarily or involuntarily files for bankruptcy or insolvency protection, or is otherwise unable to pay its debts.

Failure to pay Failure of a reference entity to make due payments greater than a specified payment requirement (commonly $1 million or more), taking into account a pre-specified grace period to prevent accidental triggering of the contract due to administrative errors.

Obligation acceleration Obligations of the reference entity have become due and payable earlier than they would have been due to default, other than a failure to pay.

Obligation default Obligations of the reference entity have become capable of being declared due and payable before they otherwise would have due to a default other than a failure to pay.

Repudiation/moratorium A reference entity or government authority rejects or challenges the validity of the obligation.

Restructuring and modified restructuring A reference entity agrees to a capital restructuring (such as a change in a loan obligation's seniority), deferral or reduction of loan, change in currency or composition of a material debt obligation such as interest or principal payments. 'Material' is generally considered to be $10 million or more. Market participants may elect an alternative definition of restructuring known as *modified restructuring* to limit the maturity and type of obligations that may be delivered by the protection seller, to reduce the 'cheapest-to-deliver' option.

APPENDIX 2.2 SAMPLE TERM SHEET FOR CREDIT DEFAULT SWAP

XYZ Bank plc
London branch

Draft Terms — Credit Default Swap

1. General Terms	
Trade Date	Aug 5, 2003
Effective Date	Aug 6, 2003
Scheduled Termination Date	Jul 30, 2005
Floating Rate Payer ('Seller')	XYZ Bank plc, London branch
Fixed Rate Payer ('Buyer')	ABC Investment Bank plc
Calculation Agent	Seller
Calculation Agent City	New York
Business Day	New York
Business Day Convention	Following
Reference Entity	Jackfruit Records Corporation
Reference Obligation	Primary Obligor: Jackfruit Records Corporation
	Maturity: Jun 30, 2020
	Coupon: 0%
	CUSIP/ISIN: xxxxx
	Original Issue Amount: USD 100,000,000
Reference Price	100%
All Guarantees	Not Applicable
2. Fixed Payments	
Fixed Rate Payer	
Calculation Amount	USD 7,000,000
Fixed Rate	0.3% per annum
Fixed Rate Payer Payment Date(s)	Oct 30, Jan 30, Apr 30, Jul 30, starting Oct 30, 2003
Fixed Rate Day Count Fraction	Actual/360

3. Floating Payments

Floating Rate Payer Calculation Amount	USD 7,000,000
Conditions to Payment	Credit Event Notice (Notifying Parties: Buyer or Seller) Notice of Publicly Available Information: Applicable (Public Source: Standard Public Sources. Specified Number: Two)
Credit Events	Bankruptcy Failure to Pay (Grace Period Extension: Not Applicable. Payment Requirement: $1,000,000)
Obligation(s)	Borrowed Money

4. Settlement Terms

Settlement Method	Physical Settlement
Settlement Currency	The currency in which the Floating Rate Payer Calculation Amount is denominated
Terms Relating to Physical Settlement	
Physical Settlement Period	The longest of the number of business days for settlement in accordance with the then-current market practice of any Deliverable Obligation being Delivered in the Portfolio, as determined by the Calculation Agent, after consultation with the parties, but in no event shall be more than 30 days
Portfolio	Exclude Accrued Interest
Deliverable Obligations	Bond or Loan
Deliverable Obligation Characteristics	Not Subordinated Specified Currency – Standard Specified Currencies Maximum Maturity: 30 years Not Contingent Not Bearer Transferable Assignable Loan Consent Required Loan
Restructuring Maturity Limitation	Not Applicable
Partial Cash Settlement of Loans	Not Applicable

Partial Cash Settlement of Assignable Loans	Not Applicable
Escrow	Applicable

5. Documentation

Confirmation to be prepared by the Seller and agreed to by the Buyer. The definitions and provisions contained in the 2003 ISDA Credit Derivatives Definitions, as published by the International Swaps and Derivatives Association, Inc., as supplemented by the May 2003 Supplement, to the 2003 ISDA Credit Derivatives Definitions (together, the 'Credit Derivatives Definitions'), are incorporated into the Confirmation

6. Notice and Account Details	
Telephone, Telex and/or Facsimile Numbers and Contact Details for Notices	Buyer: Phone: Fax: Seller: A.N. Other Phone: +1 212-xxx-xxxx Fax: +1 212-xxx-xxxx
Account Details of Seller	

Risks and Characteristics

Credit Risk. An investor's ability to collect any premium will depend on the ability of XYZ Bank plc to pay.

Non-Marketability. Swaps are not registered instruments and they do not trade on any exchange. It may be impossible for the transactor in a swap to transfer the obligations under the swap to another holder. Swaps are customized instruments and there is no central source to obtain prices from other dealers.

CHAPTER 3

Credit Derivative Instruments
Part II

We have noted that credit derivative instruments exist in funded and unfunded variants. The previous chapter looked at unfunded credit derivatives; in this chapter we consider funded credit derivatives, by which we mean principally the *credit-linked note*.

3.1 CREDIT-LINKED NOTES

Credit-linked notes (CLNs) are a form of credit derivative. They are also, in all their forms, bond instruments for which an investor pays cash, in order to receive a periodic coupon and, on maturity or termination, all or part of its initial purchase price back. That makes CLNs similar to bonds. The key difference is that the return on the CLN is explicitly linked to the credit performance of the reference security or reference entity. Like all credit derivatives, CLNs are associated with a reference entity, credit events and cash or physical settlement.

CLNs are funded credit derivatives. The buyer of the note is the investor, who is the credit-protection seller and is making an upfront payment to the protection buyer when it buys the note. Thus, the protection buyer is the issuer of the note. If no credit event occurs during the life of the note, the par redemption value of the note is paid to the investor on maturity. If a credit event does occur, then on maturity a value less than par will be paid out to the investor. This value will be reduced by the nominal value of the reference asset that the CLN is linked to. The exact process will differ according to whether *cash settlement* or *physical settlement* has been specified for the note. We will consider this later.

As with credit default swaps, CLNs are used in structured products in various combinations, although one could say that every risky cash bond is a "credit-linked note".

3.1.1 Description of CLNs

Credit-linked notes exist in a number of forms, but all of them contain a link between the return they pay and the credit-related performance of the underlying asset. A standard credit-linked note is a security, issued directly by a financial or corporate entity or by a special purpose legal entity (special purpose vehicle (SPV) or special purpose entity (SPE)), that has an interest payment and fixed maturity structure similar to a vanilla bond. The performance of the note, however, including the maturity value, is linked to the performance of a specified underlying asset or assets as well as that of the issuing entity. Notes are usually issued at par. The notes are often used by borrowers to hedge against credit risk, and by investors to enhance the yield received on their holdings. Hence, the issuer of the note is the credit protection buyer, and the buyer of the note is the credit protection seller.[1]

Essentially, credit-linked notes are hybrid instruments that combine a pure credit-risk exposure with a vanilla bond. The credit-linked note pays regular coupons; however, the credit derivative element is usually set to allow the issuer to decrease the principal amount, and/or the coupon interest, if a specified credit event occurs.

As with credit default swaps, credit-linked notes may be specified under cash settlement or physical settlement. Specifically:

- under cash settlement, if a credit event has occurred, on maturity the protection seller receives the difference between the value of the initial purchase proceeds and the value of the reference asset at the time of the credit event;
- under physical settlement, on occurrence of a credit event, the note is terminated. At maturity the protection buyer delivers the reference asset or an asset among a list of deliverable assets, and the protection seller receives the value of the original purchase proceeds minus the value of the asset that has been delivered.

Figure 3.1 illustrates a cash-settled credit-linked note.

CLNs may be issued directly by a financial or corporate entity or via an (SPV). They have been issued with the form of credit-linking taking on one or more of a number of different guises. For instance, a CLN

---

[1]Some market participants think of CLNs as being issued only by SPVs or as part of structured products. As shown in an example in this chapter though, using an example of a CLN issued direct by a corporate, this is not the case.

3.1 CREDIT-LINKED NOTES

Credit-linked note on issue

Issuer → Issue proceeds (principal payment) → Investor
Note coupons

Reference asset or entity

No credit event

Issuer → Par value on maturity (100%) → Investor

Reference asset

Credit event

Issuer → 100% minus value of reference obligation → Investor

Reference asset

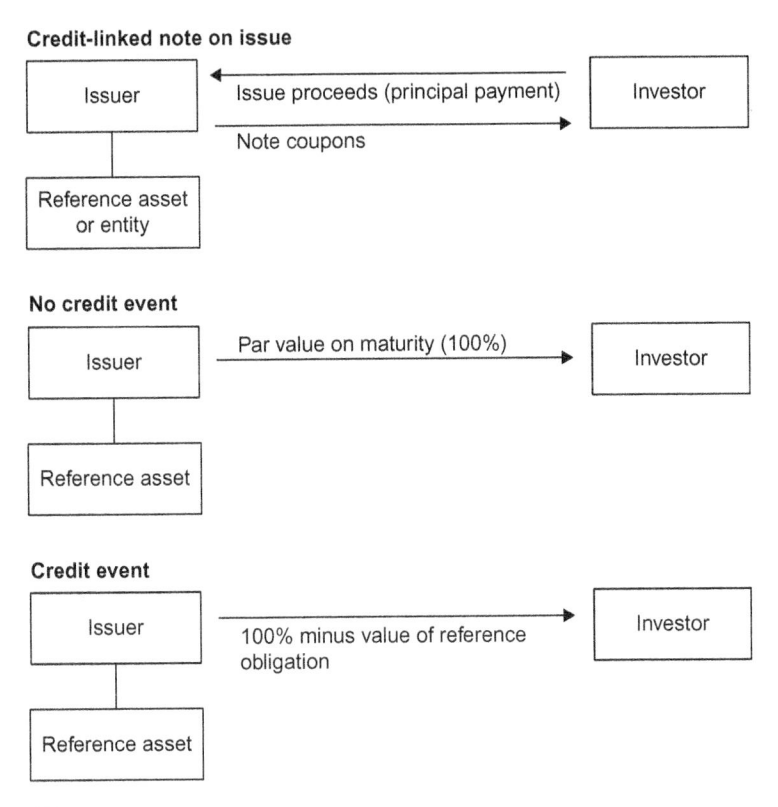

FIGURE 3.1 Credit-linked note on issue.

may have its return performance linked to the issuer's (or a specified reference entity's) credit rating, risk exposure, financial performance or circumstance of default. The return of a CLN is linked to the performance of the reference asset. The author has seen CLNs described in some texts as being 'collateralized' with the reference security, but this is incorrect. The reference security may not be owned by the issuer of the CLN, so could not possibly be described as being 'collateral' for the CLN.[2]

---

[2] Beware of this sort of mumbo-jumbo being spouted by bankers or even capital market lawyers, and in some textbooks. The proceeds of a CLN issue may be invested in collateral, as part of a collateralized synthetic obligation (CSO) structured product (see the companion book in the fixed income markets library, *Corporate bonds and structured financial products*), but in its plain vanilla form it is not collateralized by anything. It is as silly as saying that a bond issued direct by a corporate cannot be a CLN (because it has not been issued by an SPV), even though the bond may, for example, have its coupon payment linked to the credit rating of the issuer or another reference entity.

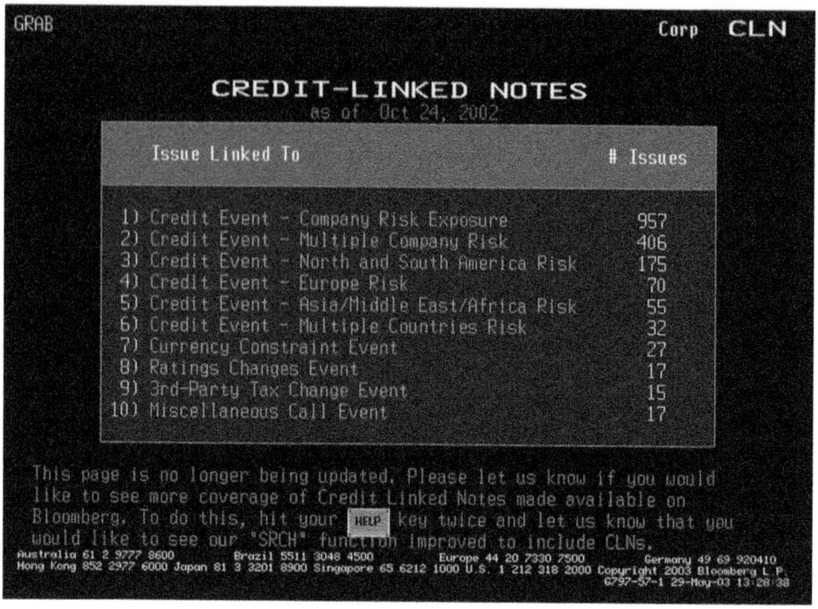

3. CREDIT DERIVATIVE INSTRUMENTS: PART II

```
GRAB                                              Corp   CLN

              CREDIT-LINKED NOTES
                  as of Oct 24, 2002

      Issue Linked To                          # Issues

      1) Credit Event - Company Risk Exposure     957
      2) Credit Event - Multiple Company Risk     406
      3) Credit Event - North and South America Risk  175
      4) Credit Event - Europe Risk                70
      5) Credit Event - Asia/Middle East/Africa Risk  55
      6) Credit Event - Multiple Countries Risk    32
      7) Currency Constraint Event                 27
      8) Ratings Changes Event                     17
      9) 3rd-Party Tax Change Event                15
     10) Miscellaneous Call Event                  17

This page is no longer being updated. Please let us know if you would
like to see more coverage of Credit Linked Notes made available on
Bloomberg. To do this, hit your HELP key twice and let us know that you
would like to see our "SRCH" function improved to include CLNs.
```

FIGURE 3.2 Bloomberg screen CLN. © *Bloomberg L.P. Reproduced with permission.*

In some texts a CLN is described as being the equivalent of a risk-free bond and a short position in a credit default swap. While this description is not incorrect, there seems no point in stating it. A CLN is a bond whose return, either principal and/or interest payments, is linked to the credit performance of a linked reference asset or reference entity, which might be the issuer. The investor in the note is selling credit protection on the reference asset or entity.

Figure 3.2 shows Bloomberg screen CLN, and a list of the various types of CLN issue that have been made. Figure 3.3 shows a page accessed from Bloomberg screen CLN that is a list of CLNs that have had their coupon affected by a change in the reference entity's credit rating, as at October 2002.

3.1.2 Illustrations

CLNs come in a variety of forms. Consider a bank issuer of credit cards that wants to fund its credit card loan portfolio via an issue of debt. The bank is rated AA−. In order to reduce the credit risk of the loans, it issues a 2-year credit-linked note. The principal amount of the bond is 100% (par) as usual, and it pays a coupon of 7.50%, which is 200 basis points above the 2-year benchmark. The equivalent spread for a vanilla bond issued by a bank of this rating would be around 120 basis points. With the CLN, though, if the incidence of bad debt amongst

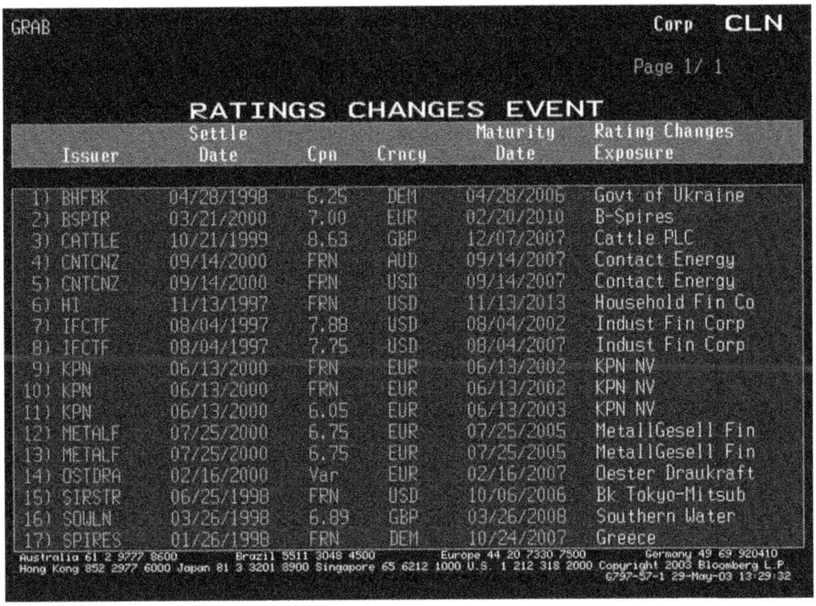

GRAB Corp CLN

Page 1/ 1

RATINGS CHANGES EVENT

Issuer	Settle Date	Cpn	Crncy	Maturity Date	Rating Changes Exposure
1) BHFBK	04/28/1998	6.25	DEM	04/28/2006	Govt of Ukraine
2) BSPIR	03/21/2000	7.00	EUR	02/20/2010	B-Spires
3) CATTLE	10/21/1999	8.63	GBP	12/07/2007	Cattle PLC
4) CNTCNZ	09/14/2000	FRN	AUD	09/14/2007	Contact Energy
5) CNTCNZ	09/14/2000	FRN	USD	09/14/2007	Contact Energy
6) HI	11/13/1997	FRN	USD	11/13/2013	Household Fin Co
7) IFCTF	08/04/1997	7.88	USD	08/04/2002	Indust Fin Corp
8) IFCTF	08/04/1997	7.75	USD	08/04/2007	Indust Fin Corp
9) KPN	06/13/2000	FRN	EUR	06/13/2002	KPN NV
10) KPN	06/13/2000	FRN	EUR	06/13/2002	KPN NV
11) KPN	06/13/2000	6.05	EUR	06/13/2003	KPN NV
12) METALF	07/25/2000	6.75	EUR	07/25/2005	MetallGesell Fin
13) METALF	07/25/2000	6.75	EUR	07/25/2005	MetallGesell Fin
14) OSTDRA	02/16/2000	Var	EUR	02/16/2007	Oester Draukraft
15) SIRSTR	06/25/1998	FRN	USD	10/06/2006	Bk Tokyo-Mitsub
16) SOWLN	03/26/1998	6.89	GBP	03/26/2008	Southern Water
17) SPIRES	01/26/1998	FRN	DEM	10/24/2007	Greece

Australia 61 2 9777 8600 Brazil 5511 3048 4500 Europe 44 20 7330 7500 Germany 49 69 920410
Hong Kong 852 2977 6000 Japan 81 3 3201 8900 Singapore 65 6212 1000 U.S. 1 212 318 2000 Copyright 2003 Bloomberg L.P.
G797-57-1 29-May-03 13:29:32

FIGURE 3.3 Bloomberg screen showing a sample of CLNs impacted by change in reference entity rating, October 2002. © *Bloomberg L.P. Reproduced with permission.*

credit card holders exceeds 10% then the terms state that note holders will only receive back \$85 per \$100 nominal. The credit card issuer has in effect purchased a credit option that lowers its liability in the event that it suffers from a specified credit event, which in this case is an above-expected incidence of bad debts. The credit card bank has issued the credit-linked note to reduce its credit exposure, in the form of this particular type of credit insurance. If the incidence of bad debts is low, the note is redeemed at par. However, if there is a high incidence of such debt, the bank will only have to repay a part of its loan liability. In this example, the reference asset linked to the CLN is the credit card loan portfolio.

Investors may wish to purchase the CLN because the coupon paid on it will be above what the credit card bank would pay on a vanilla bond it issued, and higher than other comparable investments in the market. In addition, such notes are usually priced below par on issue. Assuming the notes are eventually redeemed at par, investors will also have realized a substantial capital gain.

The majority of CLNs are issued by banks and corporates direct, in the same way as conventional bonds. An example of such a bond is shown in Figure 3.4. This shows Bloomberg screen DES for a CLN issued by British Telecom plc, the 8.125% note due in December 2010. The terms of this note state that the coupon will increase by 25 basis

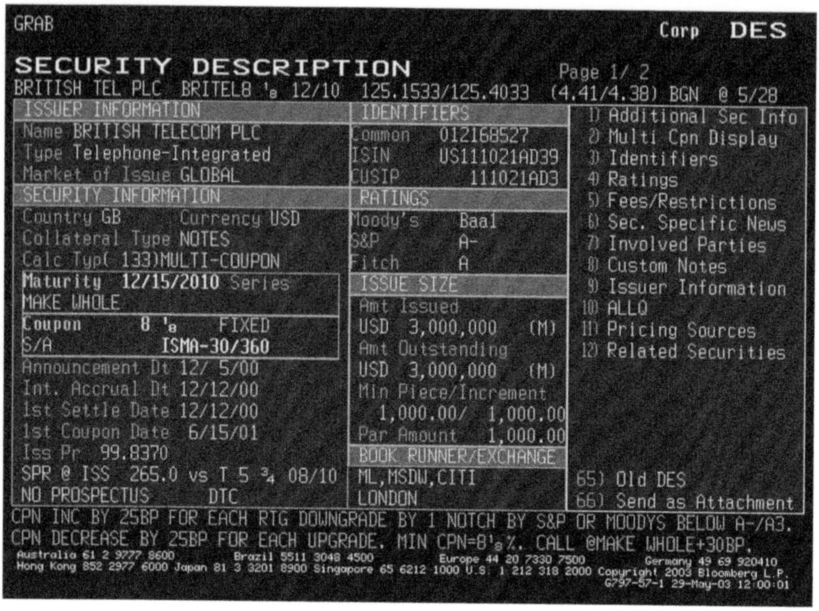

58   3. CREDIT DERIVATIVE INSTRUMENTS: PART II

```
GRAB                                                    Corp  DES
SECURITY DESCRIPTION                    Page 1/ 2
BRITISH TEL PLC  BRITEL8 ⅛ 12/10  125.1533/125.4033 (4.41/4.38) BGN @ 5/28
ISSUER INFORMATION              | IDENTIFIERS              | 1) Additional Sec Info
Name BRITISH TELECOM PLC        | Common    012168527      | 2) Multi Cpn Display
Type Telephone-Integrated       | ISIN     US111021AD39    | 3) Identifiers
Market of Issue GLOBAL          | CUSIP       111021AD3    | 4) Ratings
SECURITY INFORMATION            | RATINGS                  | 5) Fees/Restrictions
Country GB      Currency USD    | Moody's    Baa1          | 6) Sec. Specific News
Collateral Type NOTES           | S&P        A-            | 7) Involved Parties
Calc Typ( 133)MULTI-COUPON      | Fitch      A             | 8) Custom Notes
Maturity 12/15/2010 Series      | ISSUE SIZE               | 9) Issuer Information
MAKE WHOLE                      | Amt Issued               | 10) ALLQ
Coupon      8 ⅛    FIXED        | USD  3,000,000    (M)    | 11) Pricing Sources
S/A         ISMA-30/360         | Amt Outstanding          | 12) Related Securities
Announcement Dt 12/ 5/00        | USD  3,000,000    (M)
Int. Accrual Dt 12/12/00        | Min Piece/Increment
1st Settle Date 12/12/00        |   1,000.00/ 1,000.00
1st Coupon Date  6/15/01        | Par Amount    1,000.00
Iss Pr  99.8370                 | BOOK RUNNER/EXCHANGE
SPR @ ISS  265.0 vs T 5 ¾ 08/10 | ML,MSDW,CITI             | 65) Old DES
NO PROSPECTUS       DTC         | LONDON                   | 66) Send as Attachment
CPN INC BY 25BP FOR EACH RTG DOWNGRADE BY 1 NOTCH BY S&P OR MOODYS BELOW A-/A3.
CPN DECREASE BY 25BP FOR EACH UPGRADE. MIN CPN=8⅛%. CALL @MAKE WHOLE+30BP.
```

FIGURE 3.4   Bloomberg screen DES for British Telecom plc 8.125% 2010 credit-linked note issued on 5 December 2000. © *Bloomberg L.P. Reproduced with permission.*

points for each one-notch rating downgrade below A−/A3 suffered by the issuer during the life of the note. The coupon will decrease by 25 basis points for each ratings upgrade, with a minimum coupon set at 8.125%. In other words, this note allows investors to take on a credit play on the fortunes of the issuer.

Figure 3.5 shows Bloomberg screen YA for this note, as at 3 June 2003. We see that a rating downgrade meant that the coupon on the note was now 8.375%.

Figure 3.6 is the Bloomberg DES page for a USD-denominated CLN issued directly by Household Finance Corporation.[3] Like the British Telecom bond, this is a CLN whose return is linked to the credit risk of the issuer, but in a different way. The coupon of the HFC bond was issued as floating USD-Libor, but in the event of the bond not being called from November 2001, the coupon would be changed to the issuer's 2-year 'credit spread' over a fixed rate of 5.9%. In fact, the issuer called the bond with effect from the coupon change date. Figure 3.7 shows the Bloomberg screen YA for the bond and how its coupon remained as at first issue until the call date.

Another type of credit linking is evidenced from Figure 3.8. This is a JPY-denominated bond issued by Alpha-Sires, which is a medium-term

---

[3]HFC was subsequently acquired by HSBC.

3.1 CREDIT-LINKED NOTES

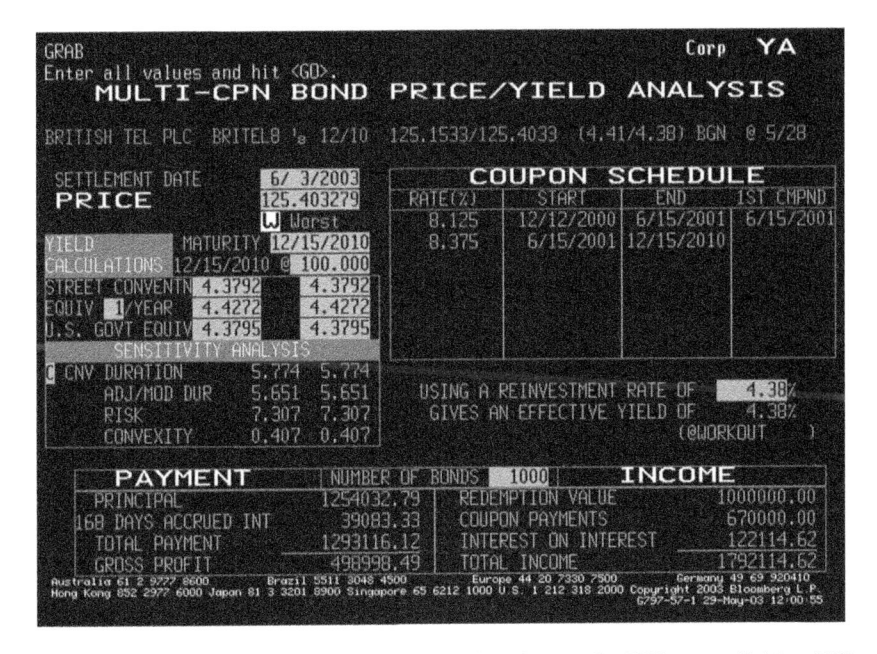

FIGURE 3.5 Bloomberg screen YA for British Telecom plc CLN, as at 12 May 2003.
© Bloomberg L.P. Reproduced with permission.

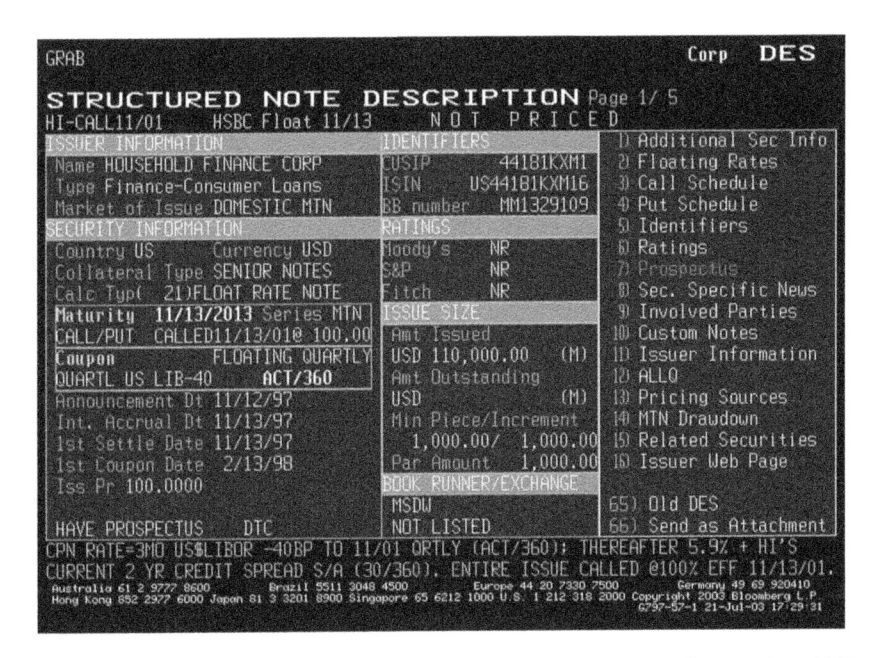

FIGURE 3.6 Bloomberg screen DES for Household Finance Corporation CLN.
© Bloomberg L.P. Reproduced with permission.

FIGURE 3.7 Bloomberg screen YA for Household Finance Corporation CLN, showing bond called (screen as at 21 July 2003). © *Bloomberg L.P. Reproduced with permission.*

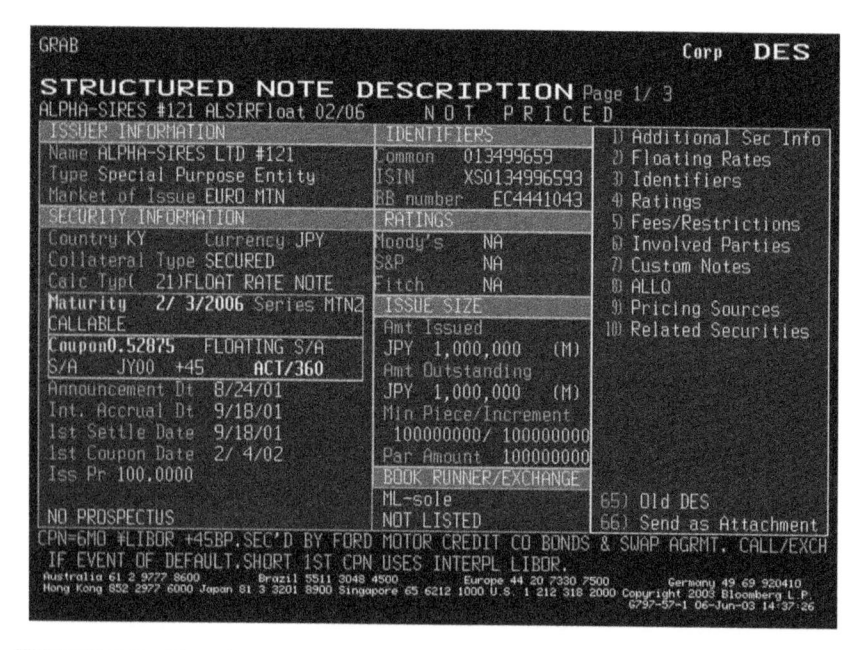

FIGURE 3.8 Bloomberg screen DES for Ford Motor credit-reference linked CLN issued by Alpha-Sires MTN programme. © *Bloomberg L.P. Reproduced with permission.*

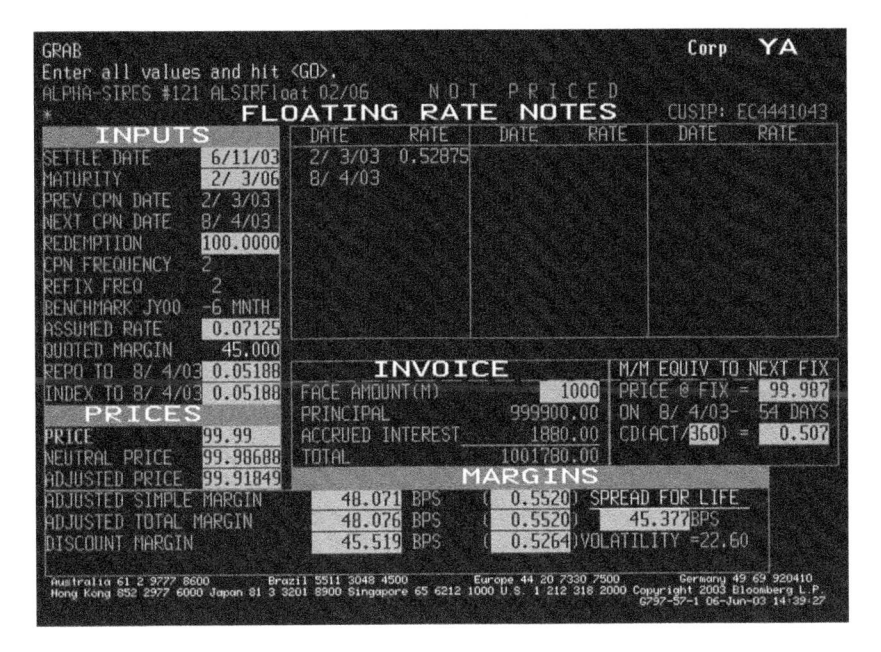

FIGURE 3.9 Bloomberg screen YA for Ford Motor credit-linked CLN as at 6 June 2003. © *Bloomberg L.P. Reproduced with permission.*

note (MTN) programme vehicle set up by Merrill Lynch. The note itself is linked to the credit quality of Ford Motor Credit. In the event of a default of the reference name, the note will be called immediately. Figure 3.9 shows the rate fixing for this note as at the last coupon date. The screen snapshot was taken on 6 June 2003.

3.2 CLNS AND STRUCTURED PRODUCTS

As with other credit derivatives, CLNs are used as part of synthetic securitization structures and structured products. We introduce this use here, while a detailed description of these products is given in the companion book in the Fixed Income Markets Library, *Corporate Bonds and Structured Finance*.

3.2.1 Simple Structure

Structured products such as synthetic collateralized debt obligations (CDOs) may combine both CLNs and credit default swaps, to meet issuer and investor requirements. For instance, Figure 3.10 shows a credit structure designed to provide a higher return for an investor on

62   3. CREDIT DERIVATIVE INSTRUMENTS: PART II

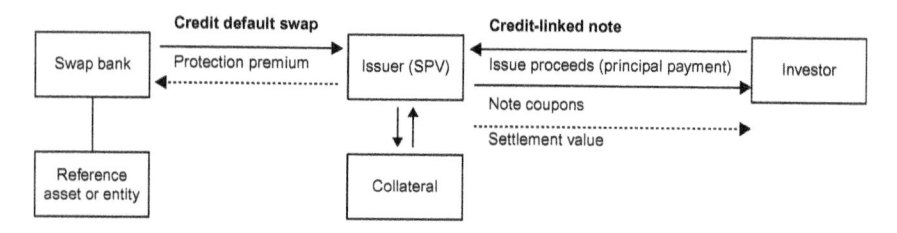

FIGURE 3.10 CLN and credit default swap structure on single reference name.

comparable risk to the cash market. An issuing entity is set up in the form of a SPV which issues CLNs to the market. The structure is engineered so that the SPV has a neutral position on a reference asset. It has bought protection on a single reference name by issuing a funded credit derivative, the CLN, and simultaneously sold protection on this name by selling a credit default swap (CDS). The proceeds of the CLN are invested in risk-free collateral, such as T-bills or a Treasury bank account. The coupon on the CLN will be a spread over Libor. It is backed by the collateral account and the fee generated by the SPV in selling protection with the credit default swap. Investors in the CLN will have exposure to the reference asset or entity, and the repayment of the note is linked to the performance of the reference entity. If a credit event occurs, the maturity date of the CLN is brought forward and the note is settled as par minus the value of the reference asset or entity.

3.2.2 The First-To-Default Credit-Linked Note

A standard credit-linked note is issued in reference to one specific bond or loan. An investor purchasing such a note is writing credit protection on a specific reference credit. A CLN that is linked to more than one reference credit is known as a *basket credit-linked note*. A development of the CLN as a structured product is the first-to-default (FtD) CLN, which is a CLN that is linked to a basket of reference assets. The investor in the CLN is selling protection on the first credit to default. Figure 3.11 shows this progression in the development of CLNs as structured products, with the *fully-funded synthetic* CDO being the vehicle that uses CLNs tied to a large basket of reference assets.

An FtD CLN is a funded credit derivative in which the investor sells protection on one reference in a basket of assets, whichever is the first to default.[4] The return on the CLN is a multiple of the average spread of the basket. The CLN will mature early on occurrence of a credit

---

[4] 'Default' here meaning a credit event as defined in the ISDA definitions.

AN INTRODUCTION TO CREDIT DERIVATIVES

3.2 CLNS AND STRUCTURED PRODUCTS

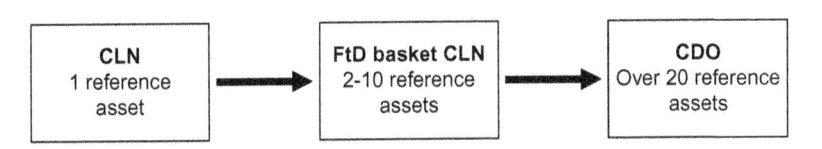

FIGURE 3.11 Progression of CLN development.

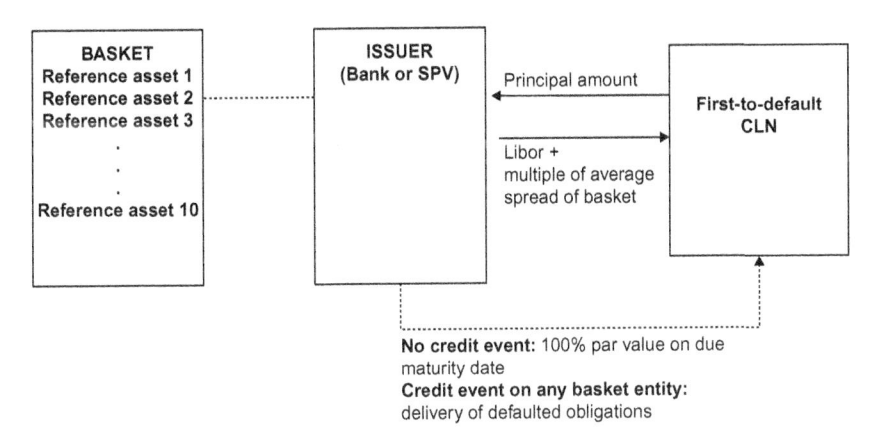

FIGURE 3.12 First-to-default CLN structure.

event relating to any of the reference assets. Note that settlement can be either of the following:

- physical settlement, with the defaulted asset(s) being delivered to the noteholder;
- cash settlement, in which the CLN issuer pays redemption proceeds to the noteholder calculated as (principal amount × reference asset recovery value).[5]

Figure 3.12 shows a generic FtD credit-linked note.

To illustrate, consider an FtD CLN issued at par with a term-to-maturity of 5 years and linked to a basket of five reference assets with a face value (issued nominal amount) of $10 million. An investor purchasing this note will pay $10 million to the issuer. If no credit event occurs during the life of the note, the investor will receive the face value of the note on maturity. If a credit event occurs on any of the assets in the basket, the note will redeem early and the issuer will deliver a

---
[5]In practice, it is not the 'recovery value' that is used but the market value of the reference asset at the time the credit event is verified. Recovery of a defaulted asset follows a legal process of administration and/or liquidation that can take some years; the final recovery value may not be known for certainty for some time.

64   3. CREDIT DERIVATIVE INSTRUMENTS: PART II

	Automobiles	Banks	Electronics	Insurance	Media	Telecoms	Utilities
AAA							
Aa1							
Aa2				SunAlliance			
Aa3		RBoS					
A1							
A2							Powergen
A3	Ford					British Telecom	
Baa1			Philips		News Intl		
Baa2							
Baa3							

FIGURE 3.13 Diversified credit exposure to basket of reference assets: hypothetical reference asset mix from 2003.

deliverable obligation of the reference entity, or a portfolio of such obligations, for a $10 million nominal amount. An FtD CLN carries a similar amount of risk exposure on default to a standard CLN, namely the recovery rate of the defaulted credit. However, its risk exposure prior to default is theoretically lower than a standard CLN, as it can reduce default probability through diversification. The investor can obtain exposure to a basket of reference entities that differ by industrial sector and by credit rating.

The matrix in Figure 3.13 illustrates how an investor can select a credit mix in the basket that diversifies risk exposure across a wide range — we show a hypothetical mix of reference assets to which an issued FtD could be linked. The precise selection of names will reflect investors' own risk/return profile requirements. It is interesting to note how these names have performed since 2003!

References

Choudhry, M. (2004a). *Corporate bonds and structured financial products*. Oxford: Butterworth-Heinemann Elsevier.

Choudhry, M. (2004b). *Structured credit products: credit derivatives and synthetic securitisation*. Singapore: John Wiley & Sons (Asia).

CHAPTER

4

Credit Derivatives: Basic Applications[1]

As derivative instruments and over-the-counter contracts, credit derivatives are very flexible products with a wide range of applications. It was their introduction that enabled synthetic structured products to be developed, which are now a major part of the debt capital markets. In the companion book in the Fixed Income Markets Library, *Corporate Bonds and Structured Financial Products*, we look in detail at synthetic securitization; in this chapter we present an overview of the basic applications of credit derivatives.

4.1 MANAGING CREDIT RISK

Credit derivatives were introduced initially as tools to hedge credit risk exposure by providing insurance against losses suffered due to 'credit events'. At market inception in 1993, commercial banks were using them to protect against losses on their corporate loan books. The principle behind credit derivatives is straightforward, and this makes them equally useful for both protection buyers and sellers. For instance, while commercial banks were offloading their loan book risk, investors who may have previously been unable to gain exposure to this sector (because of the lack of a 'market' in bank loans) could now take it on synthetically. The flexibility of credit derivatives provides users with a number of advantages precisely because they are over-the-counter

---

[1]Part of this chapter first appeared in Choudhry (2000).

4. CREDIT DERIVATIVES: BASIC APPLICATIONS

(OTC) products and can be designed to meet specific user requirements.

Banks were the first users of credit derivatives. The market developed as banks sought to protect themselves from loss due to default on portfolios of mainly illiquid assets, such as corporate loans and emerging-market syndicated loans. Whilst securitization was a well-used technique to move credit risk off the balance sheet, often this caused relationship problems with obligors, who would feel that the close relationship with their banker was being compromised if the loans were sold off the bank's balance sheet. Banks would therefore buy protection on the loan book using credit default swaps (CDSs), enabling them to hedge their credit exposure whilst maintaining banking relationships. The loan would be maintained on the balance sheet but would be protected by the CDSs.

To illustrate, consider Figure 4.1 which is a Bloomberg description page for a loan in the name of Haarman & Reimer, a chemicals company rated A3 by Moodys. We see that this loan pays 225 bps over LIBOR. Figure 4.2 shows a CDS prices page for A3-rated chemicals entities: Akzo Nobel is trading at 28 bps (to buy protection) as at 9 March 2004. A bank holding this loan can protect against default by purchasing this credit protection, and the relationship manager does not need to divulge this to the obligor.

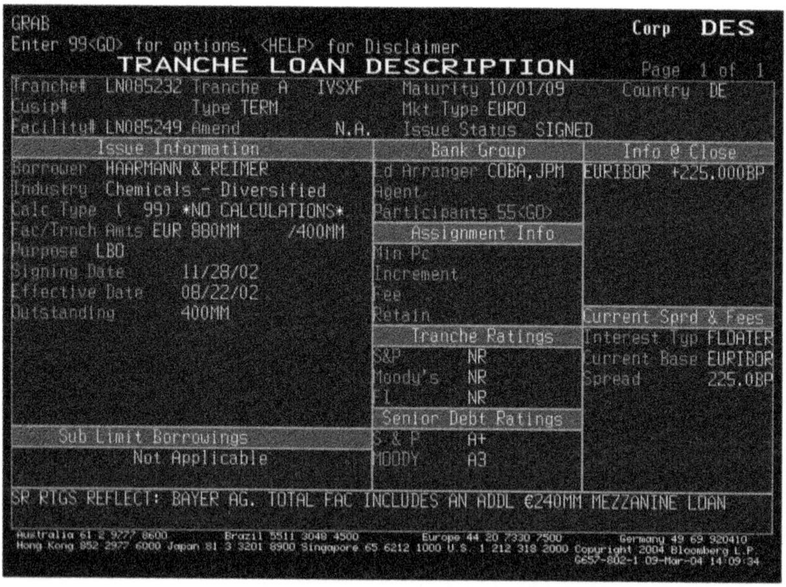

FIGURE 4.1 Haarman & Reimer loan description. © *Bloomberg LP. Reproduced with permission.*

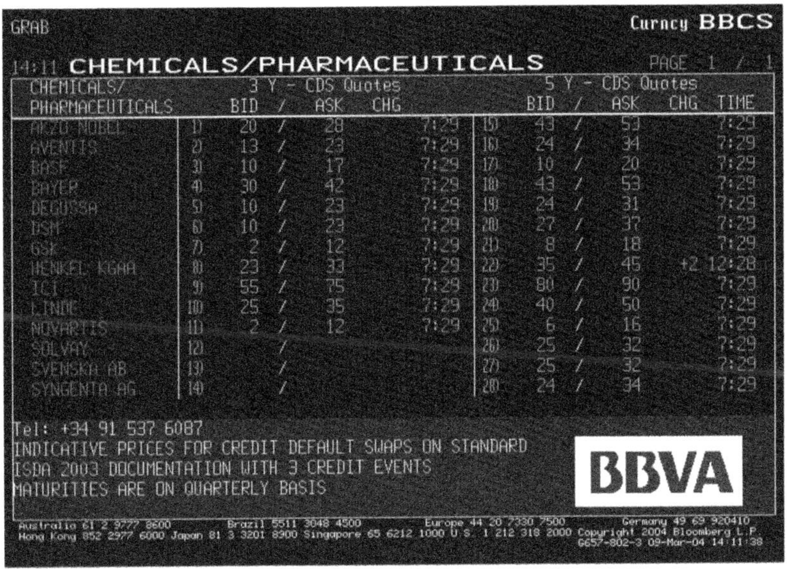

GRAB Curncy BBCS

14:11 CHEMICALS/PHARMACEUTICALS PAGE 1 / 1

CHEMICALS/ PHARMACEUTICALS		3 Y - CDS Quotes BID	/	ASK	CHG			5 Y - CDS Quotes BID	/	ASK	CHG	TIME
AKZO NOBEL	1)	20	/	28		7:29	15)	43	/	53		7:29
AVENTIS	2)	13	/	23		7:29	16)	24	/	34		7:29
BASF	3)	10	/	17		7:29	17)	10	/	20		7:29
BAYER	4)	30	/	42		7:29	18)	43	/	53		7:29
DEGUSSA	5)	10	/	23		7:29	19)	24	/	31		7:29
DSM	6)	10	/	23		7:29	20)	27	/	37		7:29
GSK	7)	2	/	12		7:29	21)	8	/	18		7:29
HENKEL KGAA	8)	23	/	33		7:29	22)	35	/	45	+2	12:28
ICI	9)	55	/	75		7:29	23)	80	/	90		7:29
LINDE	10)	25	/	35		7:29	24)	40	/	50		7:29
NOVARTIS	11)	2	/	12		7:29	25)	6	/	16		7:29
SOLVAY	12)		/				26)	25	/	32		7:29
SVENSKA AB	13)		/				27)	25	/	32		7:29
SYNGENTA AG	14)		/				28)	24	/	34		7:29

Tel: +34 91 537 6087
INDICATIVE PRICES FOR CREDIT DEFAULT SWAPS ON STANDARD
ISDA 2003 DOCUMENTATION WITH 3 CREDIT EVENTS
MATURITIES ARE ON QUARTERLY BASIS

BBVA

Australia 61 2 9777 8600 Brazil 5511 3048 4500 Europe 44 20 7330 7500 Germany 49 69 920410
Hong Kong 852 2977 6000 Japan 81 3 3201 8900 Singapore 65 6212 1000 U.S. 1 212 318 2000 Copyright 2004 Bloomberg L.P.
G657-802-3 09-Mar-04 14:11:38

FIGURE 4.2 Chemicals sector CDS prices for Banco Bilbao Vizcaya, 9 March 2004. *© Bloomberg L.P. © BBVA. Reproduced with permission.*

The other major use by banks of credit derivatives is as a product offering for clients. The CDS market has developed exactly as the market did in interest-rate swaps, with dealer banks offering two-way prices to customers and other banks as part of their product portfolio. In this role banks are both buyers and sellers of credit protection. Their net position will reflect their overall view on the market as well the other side of their customer business.

4.2 CREDIT DERIVATIVES AND RELATIVE VALUE TRADING

The existence of a liquid market in credit derivatives means that they are a viable and in some cases attractive alternative to cash market assets for investors. In some cases it is possible to exploit relative value opportunities in credit markets more efficiently by using credit derivatives, or by using a mix of cash and synthetic assets. We highlight here some approaches to credit market investment using credit derivatives. A multi-strategy investment fund will have greater freedom and flexibility to consider these applications than a more traditional long-only fund, however all credit market investment requires an expertise in credit analysis and selection, irrespective of the approach being employed.

The following strategies can be employed using either credit default swaps or cash bonds and loans, or a combination of both cash and synthetic. In some cases a specific strategy will be easier to implement using CDS.

4.2.1 Credit Selection

This is the traditional strategy that relies on picking names that are expected to 'outperform' the market. In fund management terms, one selects a diversified portfolio of credits, which ensures that systemic market risk (beta) is hedged away, while the performance of the fund generates excess return or 'alpha'. Names that are expected to outperform are trading at levels that are 'cheap' to their industry or sector class, in relative value terms. This is measured by the asset swap spread or ASW, as we observed earlier.

The fund manager selects a minimum of 10 or 12 credit-risky names that are expected to outperform other names in their sector. That is, the ASW spread for any particular name is *relatively* high compared to the sector or industry norm. Alternatively, the spread may be viewed to be higher than what the fund manager deems to be appropriate given the credit risk represented by that name. Once a name is selected, the fund manager will decide on the type of asset class (bond, loan or CDS) and its tenor. The modified duration of the asset is also relevant to the analysis because the longer the duration the more sensitive to changes in credit spreads and base interest rates the asset will be. The portfolio is hedged for currency and interest-rate risk using FX and IR derivatives. The 'credit risk' of the portfolio can be hedged using the iTraxx (or CD-X) index.

The other side of credit selection is to short names that are deemed to be 'dear' or expensive. This is problematic in the cash market as bonds have to be short-covered, and virtually impossible in the loan market. However, it is straightforward to short a name using CDS. So in this instance, the fund manager would run a portfolio of short credits, and hedge this using sold protection in the relevant iTraxx index.

4.2.2 Credit Pair Trade

In this strategy the fund manager runs a position in two different credits that attempts to exploit the spread differential between the two. As it is a relative value position, one name is put on long and the other short.[2] A pair-wise spread trade can be put on as a 'credit carry trade',

---

[2]It is possible to put this trade on in the same name, using different obligations of that name. Such a trade is more akin to an arbitrage basis trade than a true credit relative value trade, however.

4.2 CREDIT DERIVATIVES AND RELATIVE VALUE TRADING

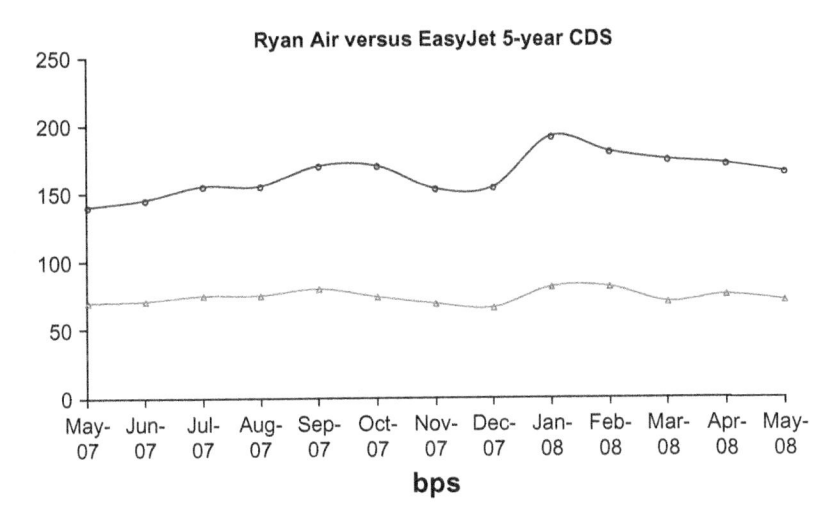

FIGURE 4.3 Ryan Air versus EasyJet Air 5-year CDS 2007–2008. *Data source: Bloomberg L.P.*

which attempts to exploit a stable spread, and a 'credit directional trade', where one expects the spread to widen or narrow. Generally the pair of credits will be names from the same industrial sector.

Consider Figure 4.3 which shows the CDS spread history between Ryan Air and EasyJet during 2007–2008. If we expect this spread to remain stable, we would buy the higher spread and sell the lower spread. This can be done using cash bonds or asset swaps, but is more straightforward using CDS. If the spread remains stable then the trade generates the long spread minus the short spread as 'carry' during the time the trade is maintained. The downside risk is if Ryan Air experiences a credit event, or if its CDS spread widens more than EasyJet's, this will create negative mark-to-market P&L. If no credit event occurs, then we have locked in the spread differential at trade inception. If Ryan Air's spread to EasyJet reduces, then this produces positive mark-to-market.

4.2.3 Basket Credit Structure Trade

In theory, a basket CDS should price at the same value as the average CDS spread of the individual constituent names, assuming each name is equally weighted. In practice there will be a difference between the theoretical price and the actual price, which can be exploited by putting on a basket credit structure. In this strategy, the investor sells protection in the basket CDS, typically a first-to-default (FtD) basket, and buys protection on each of the individual reference names in the basket.

For example, consider Figure 4.4 which shows the snapshot P&L profile for a FtD basket trade in five telecoms names, listed in Table 4.1. The investor has credit event exposure to each of the five names, although only to the first credit event. This exposure is hedged via the single-name CDS. The notional amount of the single-name CDS is a function of the delta given by each name CS01, although there is also a correlation factor to consider. The investor should assume high pairwise correlation given that all the names are from the same sector; that is, the trade is long correlation. This trade would only be put on if it generated positive carry at the start; in this example there is 277 bps positive carry based on the average spread of the five names. With the notional amounts set to produce a net zero CS01, so that the trade is credit-directional neutral, we see that it has produced an initial positive carry of just under USD 31,000.00.

From Figure 4.4 we see that the P&L profile is positive irrespective of the average spread of the five names, and a reduction in average spread increases the net profit. This is because such a reduction in the basket will not be matched to the same degree in the single-name CDS price – but this is a safe assumption only in a stable market environment.

In other words, this trade carries catastrophe risk. The net positive spread of the basket *vis-á-vis* the single names is the return demanded by an investor for assuming that the correlation value for the names is correct. If it changes, the trade will lose money. The trade carries high jump-to-default risk, the risk that one of the names suffers an almost instant deteriorating credit position, or suffers default almost overnight. As its name suggests, jump-to-default is the risk that a credit suffers a

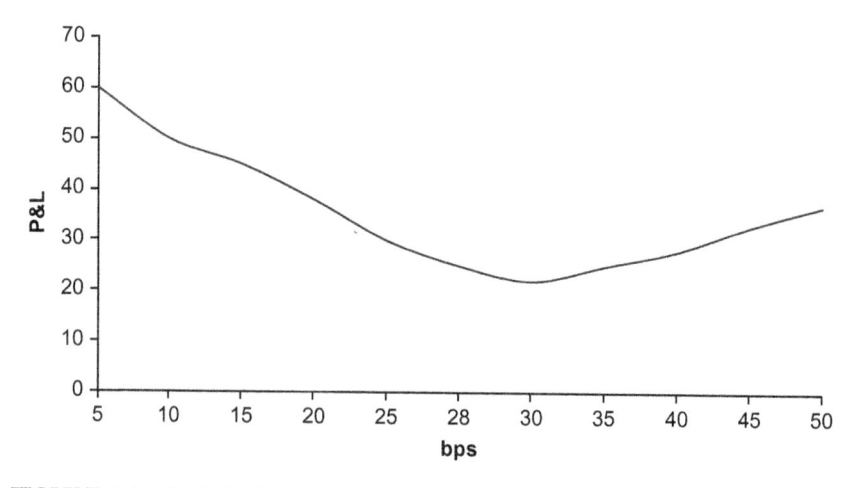

FIGURE 4.4 Credit basket trade P&L profile. *Source: Author notes.*

TABLE 4.1 CDS Basket Trade

Sell Protection	Buy Protection	Notional	Spread	CS01
FtD basket		10,000,000	461	−5,759
	British Telecom	4,800,000	201	1,158
	Vodafone	5,100,000	188	1,286
	Telefónica	4,900,000	191	1,149
	France Télécom	3,500,000	166	758
	Deutsche Telekom	5,000,000	174	1,300
	2.1459	4,660,000	920	431050
	395		184	
			277	

Source: Author notes.

credit event without showing any sign, so that its CDS price does not worsen noticeably beforehand. This is a risk to the investor because delta hedging depends on the single-name CDS positions being adjusted as their CDS spread (and thus their CS01) changes. If there is a credit event in a specific name without there being a worsening of its spread, the delta hedge in place at the time of default will be insufficient.

4.3 BOND VALUATION FROM CDS PRICES: BLOOMBERG SCREEN VCDS

Bloomberg users can use screen VCDS to obtain a CDS-price implied bond valuation, which can be compared to the actual market-observed asset swap price for a bond. This can then be used as a measure of relative value.

Figure 4.5 shows this page used to value the 9.125% of 2030 Eurobond issued by British Telecom plc as at 10 September 2008. This takes the CDS curve for the same reference name and implies a fair value that is 153 basis points below the market value (see Figure 4.6).

4.4 RELATIVE VALUE TRADING: SOVEREIGN NAMES

A relative value trade in sovereign names is the CDS-equivalent of the bond trader's yield curve relative value trade, put on to exploit changes in yield spreads. For AAA-rated sovereign names, there is little

FIGURE 4.5 Bloomberg page VCDS used to value BritTel 9.125% 2030 as at 10 September 2008. © Bloomberg L.P. All rights reserved. Visit www.bloomberg.com

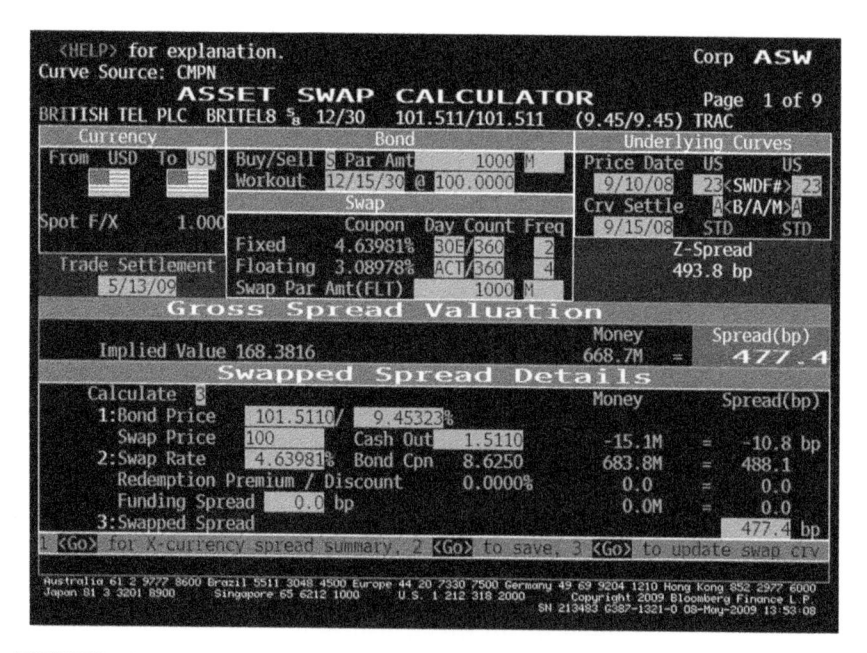

FIGURE 4.6 Bloomberg screen ASW showing asset swap spread for bond in Figure 4.5. © Bloomberg L.P. All rights reserved. Visit www.bloomberg.com

4.4 RELATIVE VALUE TRADING: SOVEREIGN NAMES

actual credit risk for the investor as we do not expect either of the reference credits to actually default. We illustrate with this example.

In the aftermath of the 2008 financial crash, CDS prices for all reference names reflected a general negative economic sentiment. A low-risk method to benefit from high prices in sovereign names, using relative value CDS trades, was an opportunity that presented itself from 2009 onwards.

4.4.1 Rationale

The market crisis has resulted in exceptional opportunities for revenue for certain sovereign names that investors would not expect to default under any circumstances. The CDS market in sovereign debt is very large and very liquid. The widening of sovereign CDS prices in countries such as Ireland and Austria was widely reported but the CDS of the strongest EU nations (Germany, France etc.) has also been dramatic. For example, to sell 5-year protection on the UK, Netherlands, France and Germany in June 2008 would have returned 11 bps, 11.3 bps, 7.4 bps and 4.9 bps respectively. Selling protection on the same countries in June 2009 paid 86.9 bps, 51.9 bps, 37.7 and 34 bps, as shown in Figure 4.7. (On 26 April 2012 these same country CDS prices were at 63.3, 124.8, 193.68 and 86.03 bps, respectively. Notice how the UK price by this time was trading below Germany, a phenomenon that resulted after the on-going Eurozone crisis from 2010.)

The trade that presented itself in 2009 was to sell protection on AAA-rated EU sovereign names to generate essentially risk-free revenue. The main risk for the trader is mark-to-market (mtm) volatility (unrealized P&L) risk. A negative mtm will generate unrealized negative P&L. To mitigate this, the investor put on an off-setting buy-protection trade in a risk-free name such as Germany sovereign name. There is a high degree of correlation between EU sovereigns and Germany, and this correlation will largely offset against mark-to-market movements on systematic risk in CDS of sovereigns. The basis differential that is exploited represents pure profit, which is realized on the unwind of the trade or on maturity of the contracts. The investor expects no default on these sovereign names during the time of the trade.

4.4.2 Example Trade Cash Flows: June 2009

In this example we sell protection on Netherlands and buy the same protection on Germany:

- Sell EUR 50 mm of protection on the Netherlands at 48 bps: we receive EUR 240,000 per year.

74   4. CREDIT DERIVATIVES: BASIC APPLICATIONS

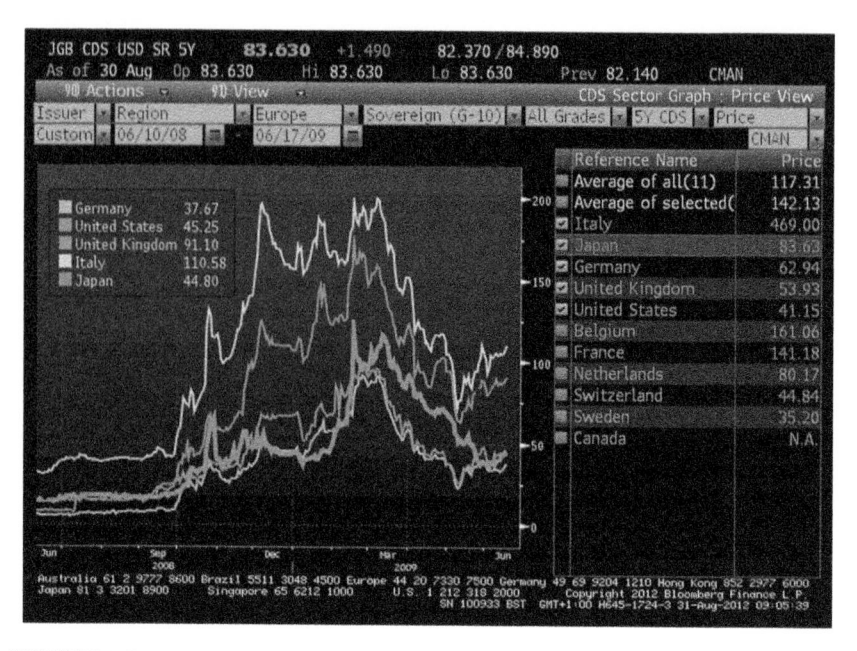

FIGURE 4.7 Sovereign CDS levels 2008–2009. © *Bloomberg L.P. Reproduced with permission. Visit www.bloomberg.com*

- Buy EUR 50 mm of protection on Germany at 31 bps: we pay EUR 155,000 per year.

Net profit per year on the above trade would be 17 bps of carry on EUR 50 mm, which is EUR 85,000.

The risk on the trade described is that the 5-year CDS on Netherlands will widen to a greater extent than Germany; this will cause mark-to-market P&L losses. There will not, however, be any loss in principle unless there is a credit event on the country in which the investor sells protection, in this case the Netherlands.

To analyze the 12-month mean spread between the two CDS values that covers the most volatile period since September 2008 (see Figure 4.8), the investor estimates potential estimated loss as the credit spread mean, which is 23.6 bps.

Under the trade described above the investor sold the spread at 17 bps so we would have lost 6.6 bps. The EUR CS01 (the EUR value of 1 bp move in the credit spread) is EUR 23,298, so using the average spread difference over 12 months as a worst case scenario the value-at-risk over the next 12 months would be:

$$6.6 \times 23,298 = \text{EUR } 153,767 \text{ mtm loss}$$

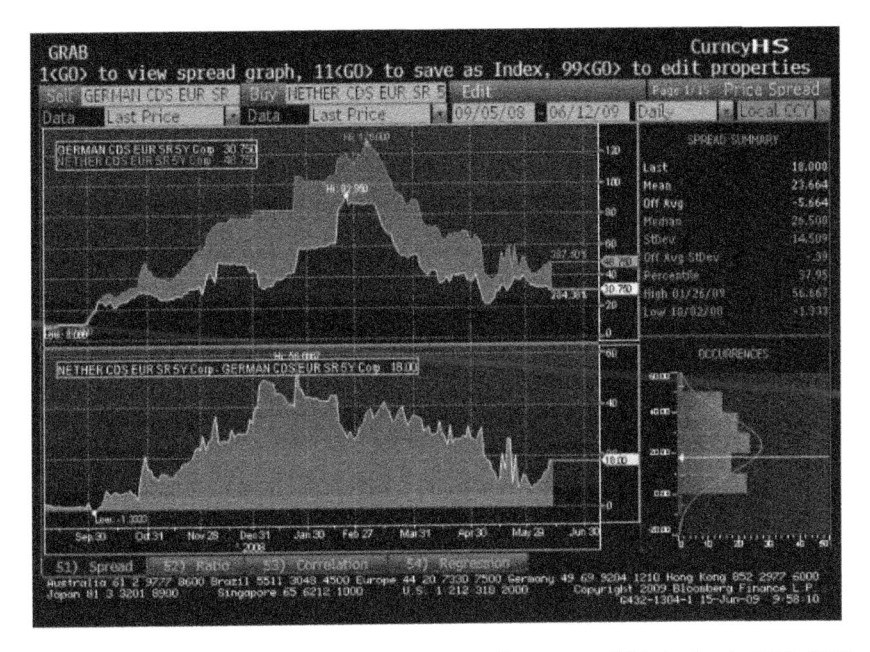

FIGURE 4.8 EUR 5-year CDS spread between Germany and Netherlands 2008–2009. © Bloomberg L.P. Reproduced with permission. Visit www.bloomberg.com

Note that this is the worst case and would be a mark-to-market unrealized loss — no realized loss would be suffered unless there was a credit event on the Netherlands.

4.5 APPLICATIONS OF TOTAL RETURN SWAPS

There are a number of reasons why portfolio managers may wish to enter into total return swap (TRS) arrangements. One of these is to reduce or remove credit risk. Using TRSs as a credit derivative instrument, a party can gain exposure to an asset without having to truly buy it. In a vanilla TRS the total return payer retains rights to the reference asset, although in some cases servicing and voting rights may be transferred. The total return receiver gains an exposure to the reference asset without having to pay out the cash proceeds that would be required to purchase it. As the maturity of the swap rarely matches that of the asset, the swap receiver may gain from the positive funding or *carry* that derives from being able to roll over short-term funding of a longer-term asset.[3] The total return payer, on the other hand, benefits from

---
[3]This assumes a positively-sloping yield curve.

protection against market and credit risk for a specified period of time, without having to liquidate the asset itself. On maturity of the swap the total return payer may reinvest the asset if it continues to own it, or it may sell the asset in the open market. Thus the instrument may be considered a synthetic repo. A TRS agreement entered into as a credit derivative is a means by which banks can take on unfunded off-balance-sheet credit exposure. Higher-rated banks that have access to Libid funding can benefit by funding on-balance-sheet assets that are credit protected through a credit derivative such as a TRS, assuming the net spread of asset income over credit protection premium is positive.

A TRS conducted as a synthetic repo is usually undertaken to effect the temporary removal of assets from the balance sheet. This may be desired for a number of reasons — for example, if the institution is due to be analyzed by credit-rating agencies or if the annual external audit is due shortly. Another reason a bank may wish temporarily to remove lower credit-quality assets from its balance sheet is if it is in danger of breaching capital limits between the quarterly return periods. In this case, as the return period approaches, lower-quality assets may be removed from the balance sheet by means of a TRS that is set to mature after the return period has passed.

We look now at some more applications of TRS instruments.

4.5.1 Capital Structure Arbitrage

A capital structure arbitrage describes an arrangement whereby investors exploit mispricing between the yields received on two different loans by the same issuer. For example, assume that the reference entity has both a commercial bank loan and a subordinated bond issue outstanding, but that the former pays Libor plus 330 basis points while the latter pays Libor plus 315 basis points. An investor enters into a total return swap in which it effectively is purchasing the bank loan and selling short the bond. The nominal amounts will be at a ratio, for argument's sake let us say 1.25:1, as the bonds will be more price-sensitive to changes in credit status than the loans.

The trade is illustrated in Figure 4.9. The investor receives the 'total return' on the bank loan, while simultaneously paying the return on

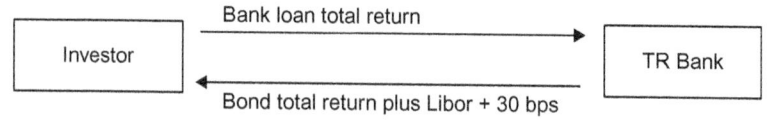

FIGURE 4.9 Total return swap in capital structure arbitrage.

the bond in addition to Libor plus 30 basis points, which is the price of the TRS. The swap generates a net spread of 67.5 basis points, given by

$$[(330 \times 1.25) - (315 + 30)]$$

4.5.2 Synthetic Repo

Total return swaps are increasingly used as synthetic repo instruments, most commonly by investors that wish to purchase the credit exposure of an asset without purchasing the asset itself. This is conceptually similar to what happened when interest-rate swaps were introduced, which enabled banks and other financial institutions to trade interest-rate risk without borrowing or lending cash funds.

Under a TRS, an asset such as a bond position may be removed from the balance sheet. In order to avoid an adverse impact on regular internal and external capital and credit exposure reporting, a bank may use TRSs to reduce the amount of lower-quality assets on the balance sheet. This can be done by entering into a short-term TRS with, say, a 2-week term that straddles the reporting date. Bonds are removed from the balance sheet if they are part of a sale plus TRS transaction. This is because legally the bank selling the asset is not required to repurchase bonds from the swap counterparty, and nor is the total return payer obliged to sell the bonds back to the counterparty (or indeed sell the bonds at all on maturity of the TRS).

Assume, then, that a portfolio manager believes that a particular bond that it does not hold is about to decline in price. To reflect this view, the portfolio manager may do one of the following:

- *Sell the bond in the market and cover the resulting short position in repo.* The cash flow out is the coupon on the bond, with capital gain if the bond falls in price. Assume that the repo rate is floating, say Libor plus a spread. The manager must be aware of the funding costs of the trade, so that unless the bond can be covered in repo at *general collateral* rates,[4] the funding will be at a loss. The yield on the bond must also be lower than the Libor plus spread received in the repo.
- *As an alternative, enter into a TRS.* The portfolio manager pays the total return on the bond and receives Libor plus a spread. If the bond yield exceeds the Libor spread, the funding will be negative; however, the trade will gain if the trader's view is proved correct

---

[4]That is, the bond cannot be *special*. A bond is special when the repo rate payable on it is significantly (say, 20–30 basis points or more) below the *general collateral* repo rate, so that covering a short position in the bond entails paying a substantial funding premium.

and the bond falls in price by a sufficient amount. If the breakeven funding cost (which the bond must exceed as it falls in value) is lower in the TRS, this method will be used rather than the repo approach. This is more likely if the bond is special.

4.5.3 The TRS as Off-Balance-Sheet Funding Tool

The TRS may be used as a funding tool, as a means of securing off-balance-sheet financing for assets held (for example) on a market making book. It is most commonly used in this capacity by broker-dealers and securities houses that have little or no access to unsecured or Libor-flat funding. When used for this purpose the TRS is similar to a repo transaction, although there are detail differences. The principal difference is that a repo is conducted between counterparties under a Global Master Repurchase Agreement (GMRA) whereas a TRS is conducted under the International Swaps and Derivatives Association (ISDA) Master standard derivative legal agreement.

Often a TRS approach is used instead of classic repo when repo is not available, or when the assets that require funding are desired to be treated as off the balance sheet for the time of the trade. At other times it may be more appropriate to use TRS for funding when the assets are less liquid or indeed not really tradeable. Bonds that are taken on by the TRS provider must be acceptable to it in terms of credit quality. If no independent price source is available the TRS provider may insist on pricing the assets itself.

As a funding tool the TRS is transacted as follows:

- The broker-dealer swaps out a bond or basket of bonds that it owns to the TRS counterparty (usually a bank), who pays the market price for the security or security.
- The maturity of the TRS can be for anything from 1 week to 1 year or even longer. For longer-dated contracts, a weekly or monthly re-set is usually employed, so that the TRS is re-priced and cashflows exchanged each week or month.
- The funds that are passed over by the TRS counterparty to the broker-dealer have the economic effect of being a loan to cover the financing of the underlying bonds. This loan is charged at Libor plus a spread.
- At the maturity of the TRS, the broker-dealer will owe interest on funds to the swap counterparty, while the swap counterparty will owe the market performance of the bonds to the broker-dealer if they have increased in price. The two cash flows are netted out.
- For a longer-dated TRS that is re-set at weekly or monthly intervals, the broker-dealer will owe the loan interest plus any decrease in

basket value to the swap counterparty at the reset date. The swap counterparty will owe any increase in value.

By entering into this transaction the broker-dealer obtains Libor-based funding for a pool of assets it already owns, while the swap counterparty earns Libor plus a spread on funds that are in effect secured by a pool of assets. This transaction takes the original assets off the balance sheet of the broker-dealer during the term of the trade, which might also be desirable.

The broker-dealer can add or remove bonds from or to the basket at each re-set date. When this happens the swap counterparty re-values the basket and will hand over more funds or receive back funds as required. Bonds are removed from the basket if they have been sold by the broker-dealer, while new acquisitions can be funded by being placed in the TRS basket.

We illustrate a funding TRS trade using an example. Figure 4.10 shows a portfolio of five hypothetical convertible bonds on the balance sheet of a broker-dealer. The spreadsheet also shows market prices. This portfolio has been swapped out to a TRS provider in a 6-month, weekly re-set TRS contract. The TRS bank has paid over the combined market value of the portfolio at a lending rate of 1.14125%. This represents 1-week Libor plus 7 basis points. We assume the broker-dealer usually funds at above this level, and that this rate is an improvement on its normal funding. However, it is not unusual for this type of trade to be undertaken even if the funding rate is not an improvement, for diversification reasons.

We see from Figure 4.10 that the portfolio has a current market value of approximately USD 151,080,000. This value is lent to the broker-dealer in return for the bonds.

One week later the TRS is re-set. We see from Figure 4.11 that the portfolio has increased in market value since the last re-set. Therefore the swap counterparty pays this difference over to the broker-dealer. This payment is netted out with the interest payment due from the broker-dealer to the swap counterparty. The interest payment is shown as USD 33,526.

Figure 4.12 shows the basket after the addition of new bonds, and the resultant change in portfolio value.

4.6 APPLICATIONS FOR PORTFOLIO MANAGERS

Credit derivatives allow market participants to separate and disaggregate credit risk, and thence to trade this risk in a secondary market. Initially portfolio managers used them to reduce credit exposure;

80   4. CREDIT DERIVATIVES: BASIC APPLICATIONS

Market Rates								
EUR/USD FX Rate	1.266550							
US$ 1W Libor	1.4055							

Name	Currency	Nominal Value	Price	Accrued	Amount	FX Rate	ISIN / CUSIP Code	Market Price	Accrued Interest
ABC Telecom	EUR	16,000,000	111.671%	0.8169%	22,795,534.57	1.2666		111.6713875	0.81693989
XYZ Bank	USD	17,000,000	128.113%	1.7472%	22,076,259.03	1.0000		128.113125	1.74722222
XTC Utility	EUR	45,000,000	102.334%	0.3135%	58,845,000.00	1.2666		102.3337875	0.31352459
SPG Corporation	EUR	30,000,000	100.32500		30,000,325.00	1.2666		100.325	0
Watty Exploited	USD	15,000,000	114.997%	0.7594%	17,363,503.13	1.0000		114.9973125	0.759375
					151,080,621.72				

Payments		
Interest ($)		
Rate	0.000000%	
Principle	151,080,000.00	
Interest Payable	+0.00	
Performance ($)		
New Portfolio Value	151,080,621.72	
Old Portfolio Value	n/a	
Performance Payment	n/a	
Net Payment ($)	+0.00	
Broker-Dealer receives from swap counterparty		

New Loan		
Portfolio Additions ($)	0.00	
New Loan Amount ($)	151,080,621.72	
New Interest Rate	1.141250%	1w Libor + 7 bps

FIGURE 4.10 Spreadsheet showing basket of bonds used in TRS funding trade.

4.6 APPLICATIONS FOR PORTFOLIO MANAGERS

EUR/USD		1.2431									
Bond	**Curr**		**Nominal Value**	**Price**	**Accrued**		**Amount**	**FX**	**ISIN/CUSIP**	**Market Price**	**Accrued**
ABC Telecom	EUR		1,60,00,000	111.50000%	0.78%		2,23,31,239	1.2431		111.5	0.77595628
XYZ Bank	USD		1,70,00,000	125.00000%	1.58%		2,15,18,931	1		125	1.58194444
XTC Utility	EUR		4,50,00,000	113.00000%	0.28%		6,33,69,825	1.2431		113	0.28278689
SPG Corporation	EUR		3,00,00,000	100.75			3,02,25,000	1.2431		100.75	
Watty Exploited	USD		1,50,00,000	113.06207%	0.63%		1,70,53,518.2	1		113.0619965	0.628125
							15,44,98,511.95				

payments
interest rate 1.14125%
amount 15,10,80,000.00   1W LIBOR + 7bps
interest payable 33,526.12   15,11,13,526.12

performance
old portfolio value 15,10,80,000.00   old portfolio value: +151,080,951.67 US$
new portfolio value 15,44,98,511.95   interest rate: 1.14125%
performance payment − 34,18,511.95   interest payable by broker-dealer +33,526.33 US$

Swap ctpy pays − 33,84,985.83 [if negative, swap counterparty pays, if positive, broker-dealer pays]   new portfolio value: +154,498,511 US$
                                  performance: 3,418,511 US$
new loan
additions —   Net payment
new loan amount 15,44,98,511.95
new interest rate 1.14875%

FIGURE 4.11 Spreadsheet showing basket of bonds at TRS re-set date plus performance and interest payments due from each TRS counterparty.

82   4. CREDIT DERIVATIVES: BASIC APPLICATIONS

EUR/USD   1.228

name	curr	Nominal	price	accrued	amount	FX	Isin	Price	Accrued
ABC Telecom	EUR	1,60,00,000	111.5000%	0.78%	2,23,31,239	1.2431		111.5	0.775956
XYZ Bank	USD	1,70,00,000	125.0000%	1.58%	2,15,18,931	1		125	1.581944
XTC Utility	EUR	4,50,00,000	113.0000000%	0.28%	6,33,69,825	1.2431		113	0.282787
SPG Corporation	EUR	3,00,00,000	100.75		3,02,25,000	1.2431		100.75	
Watty Exploited	USD	1,50,00,000	113.0620%	0.00628125	1,70,53,518	1		113.062	0.628125
Lloyd Cole Funding	USD	1,50,00,000	112.0923%	0.57%	1,68,99,628.1	1		112.0923	0.571875
					17,13,98,140.07				

payments
interest		
rate	1.14875%	1W LIBOR + 7bps
amount	15,44,98,511.95	
interest payable	34,510.03	

performance
old portfolio value	15,44,98,511.95
new portfolio value	17,13,98,140.07
performance payment	– 1,68,99,628.13
Swap cpty pays	– 1,68,65,118.09

new loan
additions	1,68,99,628.13
new loan amount	17,13,98,140.07
new interest rate	1.22750%

FIGURE 4.12 TRS basket value after addition of new bond.

subsequently they have been used in the management of portfolios, to enhance portfolio yields, and in the structuring of synthetic collateralized debt obligations. Portfolio managers' main uses of credit derivatives are summarized below.

4.6.1 Enhancing Portfolio Returns

Asset managers can derive premium income by trading credit exposures in the form of derivatives issued with synthetic structured notes. The multi-tranching aspect of structured products enables specific credit exposures (credit spreads and outright default), and their expectations, to be sold to specific areas of demand. By using structured notes, such as credit-linked notes, tied to the assets in the reference pool of the portfolio manager, the trading of credit exposures is crystallized as added yield on the asset manager's fixed income portfolio. In this way the portfolio manager has enabled other market participants to gain an exposure to the credit risk of a pool of assets but not to any other aspects of the portfolio, and without the need to hold the assets themselves.

4.6.2 Reducing Credit Exposure

Consider a portfolio manager that holds a large portfolio of bonds issued by a particular sector (say, utilities) and believes that spreads in this sector will widen in the short term. Previously, in order to reduce its credit exposure it would have to sell bonds; however, this may crystallize a mark-to-market loss and may conflict with its long-term investment strategy. An alternative approach would be to enter into a credit default swap, purchasing protection for the short term; if spreads do widen, these swaps will increase in value and may be sold at a profit in the secondary market. Alternatively, the portfolio manager may enter into total return swaps on the desired credits. It pays the counterparty the total return on the reference assets, in return for Libor. This transfers the credit exposure of the bonds to the counterparty for the term of the swap in return for the credit exposure of the counterparty.

Consider now the case of a portfolio manager wishing to mitigate credit risk from a growing portfolio (say, one that has just been launched). Figure 4.13 shows an example of an unhedged credit exposure to a hypothetical credit-risky portfolio. It illustrates the manager's expectation of credit-risk building up to $250 million as the portfolio is ramped up, and then reducing to a more stable level as the credits become more established. A 3-year CDS entered into shortly afterwards

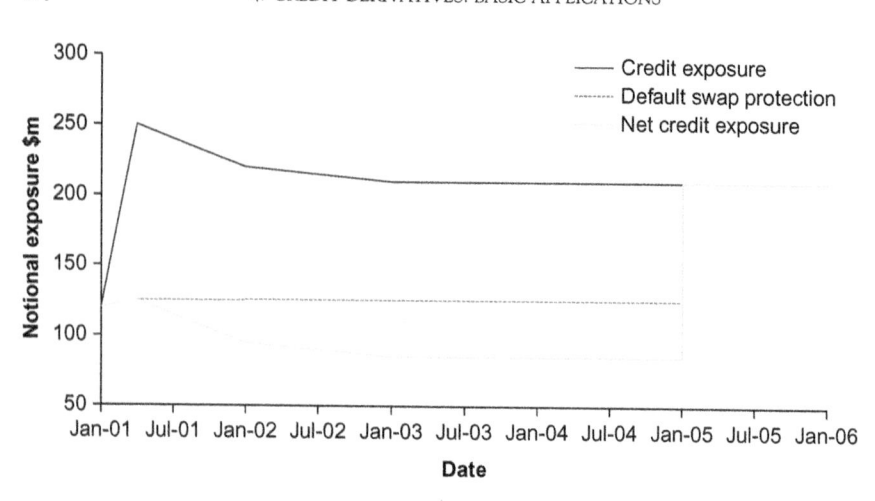

FIGURE 4.13 Reducing credit exposure.

provides protection on half of the notional exposure, shown as the broken line. The net exposure to credit events has been reduced by a significant margin.

4.6.3 Credit Switches and Zero-Cost Credit Exposure

Protection buyers utilizing CDS must pay a premium in return for laying off their credit-risk exposure. An alternative approach for an asset manager involves the use of credit switches for specific sectors of the portfolio. In a credit switch, the portfolio manager purchases credit protection on one reference asset or pool of assets and simultaneously sells protection on another asset or pool of assets.[5] So, for example, the portfolio manager would purchase protection for a particular fund and sell protection on another. Typically the entire transaction would be undertaken with one investment bank, which would price the structure so that the net cash flows would be zero. This has the effect of synthetically diversifying the credit exposure of the portfolio manager, enabling it to gain and/or reduce exposure to sectors as it desires.

4.6.4 Exposure to Market Sectors

Investors can use credit derivatives to gain exposure to sectors for which they do not wish a cash-market exposure. This can be achieved with an *index* swap, which is similar to a TRS, with one counterparty

---

[5] A pool of assets would be concentrated on one sector, such as utility company bonds.

paying a total return that is linked to an external reference index. The other party pays a Libor-linked coupon or the total return of another index. Indices that are used might include the government bond index, a high-yield index or a technology stocks index. Assume that an investor believes that the bank loan market will outperform the mortgage-backed bond sector; to reflect this view the investor enters into an index swap in which he pays the total return of the mortgage index and receives the total return of the bank loan index.

Another possibility is synthetic exposure to foreign currency and money markets. Again, we assume that an investor has a particular view on an emerging market currency. If he wishes he can purchase a short-term (say 1-year) domestic coupon-bearing note, whose principal redemption is linked to a currency factor. This factor is based on the ratio of the spot value of the foreign currency on issue of the note to the value on maturity. Such currency-linked notes can also be structured so that they provide an exposure to sovereign credit risk. The downside of currency-linked notes is that if the exchange rate goes the other way, the note will have a zero return – in effect a negative return once the investor's funding costs have been taken into account.

4.6.5 Credit Spreads

Credit derivatives can be used to trade credit spreads. Assume that an investor has negative views on a certain emerging market government bond credit spread relative to UK gilts. The simplest way to reflect this view would be to go long on a credit default swap on the sovereign, paying X basis points. Assuming that the investor's view is correct and the sovereign bonds decrease in price as their credit spread widens, the premium payable on the credit swap will increase. The investor's swap can then be sold into the market at this higher premium.

Reference

Choudhry, M. (2000). *Credit derivatives: an introduction for portfolio managers*. YieldCurve.com, December 2000.

CHAPTER 5

Credit Derivatives Pricing and Valuation

In this chapter we look at the various approaches used in pricing and valuation of credit derivatives. We consider generic techniques and compare prices obtained using different pricing models. We also present a first-principles method of pricing credit default swaps based on the original pricing approach articulated by Professor Darrell Duffie in 1992.

5.1 INTRODUCTION

The pricing of credit derivatives should aim to provide a 'fair value' for the credit derivative instrument. In the sections below we discuss the pricing models currently used in the industry. The effective use of pricing models requires an understanding of the models' assumptions and the key pricing parameters, and a clear understanding of the limitations of a pricing model. Issues to consider when carrying out credit derivative pricing include:

- implementation and selection of appropriate modelling techniques;
- parameter estimation;
- quality and quantity of data to support parameters and calibration;
- calibration to market instruments for risky debt.

For credit derivative contracts in which the payout is on credit events other than default, the modelling of the credit evolutionary path is critical. If, however, a credit derivative contract does not pay out on intermediate stages between the current state and default, then the important factor is the probability of default from the current state.

Before continuing with this chapter, readers may wish to look at the section that discusses asset swap pricing methods, part of our discussion on the *basis*, in Chapter 7. This was commonly used at the inception of the credit derivatives market, but is not used today due to the inherent differences between asset swaps and other credit derivatives. But understanding this approach helps with an appreciation of credit default swap (CDS) pricing.

We now consider a number of pricing models as used in the credit derivative markets.

5.2 PRICING MODELS

Pricing models for credit derivatives fall into two classes:

- structural models;
- reduced form models.

We discuss these models below.

5.2.1 Structural Models

Structural models are characterized by modelling the firm's value in order to provide the probability of a firm default. The Black–Scholes–Merton option pricing framework is the foundation of the structural model approach. The default event is assumed to occur when the firm's assets fall below the book value of the debt.

Merton applied option pricing techniques to the valuation of corporate debt (Merton, 1974). By extension, the pricing of credit derivatives based on corporate debt may in some circumstances be treated as an option on debt (which is therefore analogous to an option on an option model).

Merton models have the following features:

- default events occur predictably when a firm has insufficient assets to pay its debt;
- a firm's assets evolve randomly, and the probability of a firm default is determined using the Black–Scholes–Merton option pricing theory.

Some practitioners argue that Merton models are more appropriate than reduced form models when pricing default swaps on high-yield bonds, due to the higher correlation of high-yield bonds with the underlying equity of the issuer firm.

The constraint of structural models is that the behaviour of the value of assets and the parameters used to describe the process for the value of the firm's assets are not directly observable, and the method does not consider the underlying market information for credit instruments.

5.2.2 Reduced Form Models

Reduced form models are a form of no-arbitrage model. These models can be fitted to the current term structure of risky bonds to generate no-arbitrage prices. In this way, the pricing of credit derivatives using these models will be consistent with the market data on the credit-risky bonds traded in the market. These models allow the default process to be separated from the asset value, and are more commonly used to price credit derivatives.

Some key features of reduced form models include:

- complete and arbitrage-free credit market conditions are assumed;
- recovery rate is an input into the pricing model;
- use of credit spread data to estimate the risk-neutral probabilities;
- use of transition probabilities from credit agencies can be accommodated in some of these models — the formation of the risk-neutral transition matrix from the historical transition matrix is a key step;
- default can take place randomly over time and the default probability can be determined using the risk-neutral transition matrix.

When implementing reduced form models it is necessary to consider issues such as the illiquidity of underlying credit-risky assets. Liquidity is often assumed to be present when we develop pricing models; however, in practice there may be problems when calibrating a model to illiquid positions, and in such cases the resulting pricing framework may be unstable and provide the user with spurious results. Another issue is the relevance of using historical credit transition data, used to project future credit migration probabilities. In practice it is worthwhile reviewing the sensitivity of price to the historical credit transition data when using the model.

The key reduced form models that provide a detailed modelling of default risk include those presented by Jarrow and Lando (1997), Das and Tufano (1996) and Duffle and Singleton (1997). We consider these models in this section.

Jarrow, Lando and Turnbull (JLT) Model

This model focuses on modelling default and credit migration. Its data and assumptions include the use of:

- a statistical rating transition matrix that is based on historic data;
- risky bond prices from the market used in the calibration process;
- a constant recovery rate assumption (the recovery amount is assumed to be received at the maturity of the bond);
- a credit spread assumption for each rating level.

It also assumes no correlation between interest rates and credit rating migration.

The statistical transition matrix is adjusted by calibrating the expected risky bond values to the market values for risky bonds. The adjusted matrix is referred to as the risk-neutral transition matrix. The risk-neutral transition matrix is key to the pricing of several credit derivatives.

The JLT model allows the pricing of default swaps, as the risk-neutral transition matrix can be used to determine the probability of default. The JLT model is sensitive to the level of the recovery rate assumption and the statistical rating matrix. It has a number of advantages; as the model is based on credit migration, it allows the pricing of derivatives for which the payout depends on such credit migration. In addition, the default probability can be explicitly determined and may be used in the pricing of credit default swaps.

The disadvantages of the model include the fact that it depends on the selected historical transition matrix. The applicability of this matrix to future periods needs to be considered carefully – whether, for example, it adequately describes future credit migration patterns. In addition, it assumes all securities with the same credit rating have the same spread, which is restrictive. For this reason, the spread levels chosen in the model are a key assumption in the pricing model. Finally, the constant recovery rate is another practical constraint, as in practice the level of recovery will vary.

Das–Tufano Model

The Das–Tufano (DT) model is an extension of the JLT model. The model aims to produce the risk-neutral transition matrix in a similar way to the JLT model; however, this model uses stochastic recovery rates. The final risk-neutral transition matrix should be computed from the observable term structures. The stochastic recovery rates introduce more variability in the spread volatility. Spreads are a function of factors which may not only be dependent on the rating level of the credit, as in practice credit spreads may change even though credit ratings have not changed. Therefore, to some extent the DT model introduces this additional variability into the risk-neutral transition matrix.

Various credit derivatives may be priced using this model; for example, credit default swaps, total return swaps and credit spread options. The pricing of these products requires the generation of the appropriate credit-dependent cash flows at each node on a lattice of possible outcomes. The fair value may be determined by discounting the probability-weighted cash flows. The probability of the outcomes would be determined by reference to the risk-neutral transition matrix.

5.2 PRICING MODELS

Duffie–Singleton Approach

The Duffie–Singleton modelling approach considers the three components of risk for a credit-risky product, namely the risk-free rate, the hazard rate, and the recovery rate.

The *hazard rate* characterizes the instantaneous probability of default of the credit-risky underlying exposure. As each of the components above may not be static over time and a pricing model may assume a process for each of these components of risk, the process may be implemented using a lattice approach for each component. The constraint on the lattice formation is that this lattice framework should agree to the market pricing of credit-risky debt.

Here we demonstrate that the credit spread is related to risk of default (as represented by the hazard rate) and the level of recovery of the bond. We assume that a zero-coupon risky bond maturing in a small time element Δt where:

- λ is the annualized hazard rate
- φ is the recovery value
- r is the risk-free rate
- s is the credit spread

and where its price P is given by:

$$P = e^{-r\Delta t}((1 - \lambda \Delta t) + (\lambda \Delta t)\varphi) \quad (5.1)$$

Alternatively P may be expressed as:

$$P \cong e^{-\Delta t (r + \lambda)(1 - \varphi)} \quad (5.2)$$

However, as the usual form for a risky zero-coupon bond is:

$$P = e^{-\Delta t(r+s)} \quad (5.3)$$

Therefore we have shown that:

$$s \cong \lambda(1 - \varphi) \quad (5.4)$$

This would imply that the credit spread is closely related to the hazard rate (that is, the likelihood of default) and the recovery rate.

This relationship between the credit spread, the hazard rate and recovery rate is intuitively appealing. The credit spread is perceived to be the extra yield (or return) an investor requires for credit risk assumed. For example:

- as the hazard rate (or instantaneous probability of default) rises then the credit spread increases;
- as the recovery rate decreases the credit spread increases.

A 'hazard rate' function may be determined from the term structure of credit. The hazard rate function has its foundation in statistics, and may be linked to the instantaneous default probability.

The hazard rate function ($\lambda(s)$) can then be used to derive a probability function for the survival function $S(t)$:

$$S(t) = \exp\left[-\int_0^t \lambda(s)ds\right] \qquad (5.5)$$

The hazard rate function may be determined by using the prices of risky bonds. The lattice for the evolution of the hazard rate should be consistent with the hazard rate function implied from market data. An issue when performing this calibration is the volume of relevant data available for the credit.

Recovery Rates

The recovery rate usually takes the form of the percentage of the par value of the security recovered by the investor.

The key elements of the recovery rate include:

- the level of the recovery rate;
- the uncertainty of the recovery rate based on current conditions specific to the reference credit;
- the time interval between default and the recovery value being realized.

Generally, recovery rates are related to the seniority of the debt. Therefore if the seniority of debt changes then the recovery value of the debt may change. Also, recovery rates exhibit significant volatility.

5.3 CREDIT SPREAD MODELLING

Although spreads may be viewed as a function of default risk and recovery risk, spread models do not attempt to break down the spread into its default risk and recovery risk components.

The pricing of credit derivatives that pay out according to the level of the credit spread would require that the credit spread process is adequately modelled. In order to achieve this, a stochastic process for the distribution of outcomes for the credit spread is an important consideration.

5.3 CREDIT SPREAD MODELLING

An example of the stochastic process for modelling credit spreads, which may be assumed, includes a mean reverting process such as:

$$ds = \kappa(\mu - s)dt + \sigma s dw \tag{5.6}$$

where

- ds is the change in the value of the spread over an element of time (dt)
- dt is the element of time over which the change in spread is modelled
- s is the credit spread
- κ is the rate of mean reversion
- μ is the mean level of the spread
- dw is the Wiener increment (sometimes written dW, dZ or dz)
- σ is the volatility of the credit spread

In this model, when s rises above a mean level of the spread the drift term $(\mu - s)$ will become negative and the spread process will drift towards (revert) to the mean level. The rate of this drift towards the mean is dependent on κ, the rate of mean reversion.

The pricing of a European spread option requires the distribution of the credit spread at the maturity (T) of the option. The choice of model affects the probability assigned to each outcome. The mean reversion factor reflects the historic economic features over time of credit spreads, to revert to the average spreads after larger than expected movements away from the average spread.

Therefore the European option price may be reflected as:

$$\text{Option price} = E[e^{-rT}(\text{Payoff}(s, X))] = e^{-rT} \int_0^\infty f(s, X) p(s) ds \tag{5.7}$$

where:

- X is the strike price of the spread option
- $p(s)$ is the probability function of the credit spread
- $E[\ ]$ denotes the expected value
- $f(s, X)$ is the payoff function at maturity of the credit spread

More complex models for the credit spread process may take into account factors such as the term structure of credit and possible correlation between the spread process and the interest process.

The pricing of a spread option is dependent on the underlying process. As an example we compare the pricing results for a spread option model including mean reversion to the pricing results from a standard Black–Scholes model in Tables 5.1 and 5.2.

Tables 5.1 and 5.2 show the sensitivity on the pricing of a spread option to changes to the underlying process. Comparing Tables 5.1 and

5.2 shows the impact of the time to expiry increasing by 6 months. In a mean reversion model the mean level and the rate of mean reversion are important parameters that may significantly affect the probability distribution of outcomes for the credit spread, and hence the price.

5.4 PRODUCT PRICING APPROACH

5.4.1 The Forward Credit Spread

The forward credit spread can be determined by considering the spot prices for the risky security and risk-free benchmark security, while the forward yield can be derived from the forward price of these securities. The forward credit spread is the difference between the forward risky security yield and the forward yield on a risk-free security. The forward credit spread is calculated by using yields to the forward date and the yield to the maturity of the risky assets.

TABLE 5.1 Comparison of Model Results, Expiry in 6 Months

Expiry in 6 months Risk-free rate = 10% Strike = 70 bps Credit spread = 60 bps Volatility = 20%	Mean Reversion Model Price	Standard Black–Scholes Price	% Difference Between Standard Black–Scholes and Mean Reversion Model Price
Mean level = 50 bps $K = .2$			
Put	0.4696	0.5524	17.63
Call	10.9355	9.7663	11.97
Mean level = 50 bps $K = .3$			
Put	0.3510	0.5524	57.79
Call	11.2031	9.7663	14.12
Mean level = 80 bps $K = .2$			
Put	0.8729	0.5524	58.02
Call	8.4907	9.7663	15.02
Mean level = 80 bps $K = .3$			
Put	0.8887	0.5524	60.87
Call	7.5411	9.7663	29.51

5.4 PRODUCT PRICING APPROACH

TABLE 5.2 Comparison of Model Results, Expiry in 12 Months

Expiry in 12 Months Risk-Free Rate = 10% Strike = 70 bps Credit Spread = 60 bps Volatility = 20%	Mean Reversion Model Price	Standard Black–Scholes Price	% Difference Between Standard Black–Scholes and Mean Reversion Model Price
Mean level = 50 bps K = .2			
Put	0.8501	1.4331	68.58
Call	11.2952	10.4040	8.56
Mean level = 50 bps K = .3			
Put	0.7624	1.4331	87.97
Call	12.0504	10.4040	15.82
Mean level = 80 bps K = .2			
Put	1.9876	1.4331	38.69
Call	7.6776	10.4040	35.51
Mean level = 80 bps K = .3			
Put	2.4198	1.4331	68.85
Call	6.7290	10.4040	54.61

For example, the following data are used in determining the forward credit spread:

Current date	1/2/98
Forward date	1/8/98
Maturity	1/8/06
Time period from current date to maturity:	8 years and 6 months
Time period from current date to forward date:	6 months
Yield to forward date:	
Risk-free security	6.25%
Risky security	6.50%
Yield to maturity:	
Risk-free security	7.80%
Risky security	8.20%

The forward yields (calculated from inputs above; see below for detailed derivation) are:

Risk-free security   7.8976%

Risky security   8.3071%

The details of the calculation of forward rates are as follows.
Risk-free security:

$$(1.0780)^{86/12} = (1.0625)^{6/12} * (1+rf_{riskfree})^8$$

where $rf_{riskfree}$ is the forward risk-free rate implied by the yields on a risk-free security. This equation implies that $rf_{riskfree}$ is 7.8976%.
Similarly for the risky security we have:

$$(1.082)^{86/12} = (1.0625)^{6/12} * (1+rf_{risky})^8$$

where rf_{risky} is the forward risky rate implied by the yields on a risky security. This equation implies that rf_{risky} is 8.3071%.

Therefore the forward credit spread is the difference between the forward rate implied by the risky security less the forward rate implied by the yields on a risk-free security. In the example above, this is

$$rf_{risky} - rf_{riskfree} = 8.3071 - 7.8979 = 0.4095\%$$

The current spread is equal to $8.20 - 7.80 = 0.40\% = 40$ bps.
The difference between the forward credit spread and the current spread is $0.4095 - 0.40 = 0.0095\% = 0.95$ basis points.
The calculation of the forward credit spread is critical to the valuation of credit spread products as the payoff of spread forwards is highly sensitive to the implied forward credit spread.

5.4.2 Asset Swaps Pricing

Assume that an investor holds a bond and enters into an asset swap with a bank. Then the value of an asset swap is the spread the bank pays over or under Libor. This is based on the following components:

- value of the coupons of the underlying asset compared to the market swap rate;
- the accrued interest and the clean price premium or discount compared to par value. Thus when pricing the asset swap it is necessary to compare the par value to the underlying bond price.

The spread above or below Libor reflects the credit spread difference between the bond and the swap rate.

The Bloomberg asset swap calculator pricing screens in Chapters 2 and 3 show these components in the analysis of the swapped spread details.

For example, let us assume that we have a credit-risky bond with the following details:

Currency:	EUR
Issue date:	31 March 2000
Maturity:	31 March 2007
Coupon:	5.5% per annum
Price (dirty):	105.3%
Price (clean):	101.2%
Yield:	5%
Accrued interest:	4.1%
Rating:	A1

To buy this bond, the investor would pay 105.3% of par value. The investor would receive the fixed coupons of 5.5% of par value. Let us assume that the swap rate is 5%. The investor in this bond enters into an asset swap with a bank in which the investor pays the fixed coupon and receives Libor ± spread.

The asset swap price (i.e. spread) on this bond has the following components:

- The value of the excess value of the fixed coupons over the market swap rate is paid to the investor. Let us assume that in this case this is approximately 0.5% when spread into payments over the life of the asset swap.
- The difference between the bond price and par value is another factor in the pricing of an asset swap. In this case the price premium which is expressed in present value terms should be spread over the term of the swap and treated as a payment by the investor to the bank (if a

dirty price is at a discount to the par value, then the payment is made from the bank to the investor). For example, in this case let us assume that this results in a payment from the investor to the bank of approximately 0.23% when spread over the term of the swap.

These two elements result in a net spread of 0.5% − 0.23% = 0.27%. Therefore, the asset swap would be quoted as Libor + 0.27% (or Libor plus 27 bps).

5.4.3 Total Return Swap (TRS) Pricing

The present value of the two legs of the TRS should be equivalent. This would imply that the level of the spread is therefore dependent on the following factors:

- credit quality of the underlying asset;
- credit quality of the TRS counterparty;
- capital costs and target profit margins;
- funding costs of the TRS provider, as it will hedge the swap by holding the position in the underlying asset.

The fair value for the TRS will be the value of the spread for which the present value of the Libor ± spread leg equals the present value of the returns on the underlying reference asset. The present value of the returns on the underlying reference asset may be determined by evolving the underlying reference asset. The expected value of the TRS payoff at maturity should be discounted to the valuation date.

The reduced form models described earlier are a new generation of credit derivative pricing models, which are now increasingly being used to price total return swaps.

5.5 CREDIT CURVES

The credit curves (or default swap curves) reflect the term structure of spreads by maturity (or tenor) in the credit default swap markets. The shape of the credit curves is influenced by the demand and supply for credit protection in the credit default swaps market and reflects the credit quality of the reference entities (both specific and systematic risk). The changing levels of credit curves provide traders and arbitragers with the opportunity to measure relative value and establish credit positions.

In this way, any changes of shape and perceptions of the premium for CDS protection are reflected in the spreads observed in the market. In periods of extreme price volatility, as seen in the middle of 2002, the curves may invert to reflect the fact that the cost of protection for

5.5 CREDIT CURVES

shorter-dated protection trades at wider levels than the longer-dated protection. This is consistent with the pricing theory for credit default swaps.

The probability of survival for a credit may be viewed as a decreasing function against time. The survival probabilities for each traded reference credit can be derived from its credit curve. The survival probability is a decreasing function, because it reflects the fact that the probability of survival for a credit reduces over time – for example, the probability of survival to year 3 is higher than the probability of survival to year 5.

Under non-volatile market conditions, the shape of the survival probability and the resulting credit curve will take a different form to the shape implied in volatile market conditions; the graphs may change to reflect the higher perceived likelihood of default. For example, the shape of the survival probability may take the form as shown in Figure 5.1.

The corresponding credit curves consistent with these survival probabilities take the form shown in Figure 5.2.

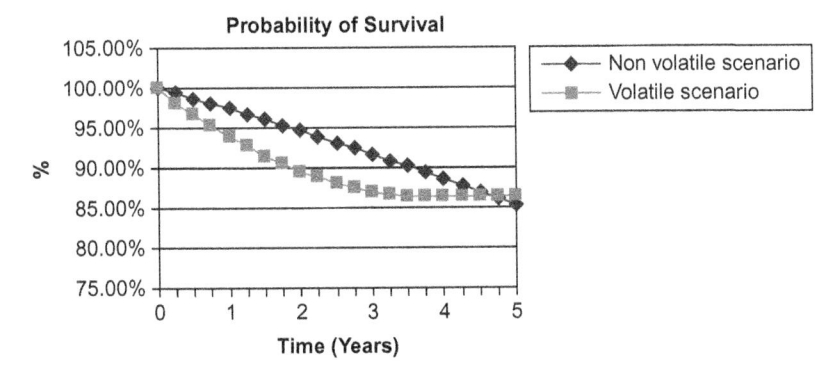

FIGURE 5.1 Probability of survival.

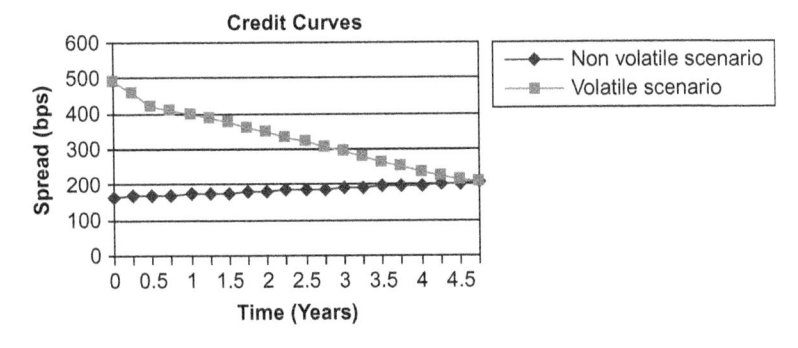

FIGURE 5.2 Credit curves.

AN INTRODUCTION TO CREDIT DERIVATIVES

This shows that the credit curve inversion is consistent with the changes in the survival probability functions.

In this analysis, we assume that the assumed recovery rate for the 'cheapest to deliver' bond remains the same at 35% of notional value.

References

Das, S., & Tufano, P. (1996). Pricing credit-sensitive debt when interest rate, credit ratings and credit spreads are stochastic. *Journal of Finance Engineering, 7,* 133–149.

Duffle, D., & Singleton, K. (1997). Modelling term structures of defaultable bonds. *Review of Financial Studies, 11,* 61–74.

Jarrow, R., & Lando, D. (1997). A Markov model for the term structure of credit spreads. *Review of Financial Studies, 10,* 481–523.

Merton, R. C. (1974). On the pricing of corporate debt: the risk structure of interest rates. *Journal of Finance, 29,* 449–470.

CHAPTER 6

Credit Default Swap Pricing

We concentrate specifically now on the credit default swap (CDS), and a market approach for pricing these instruments. We consider here the plain vanilla structure, in which a protection buyer pays a regular premium to a protection seller, up to the maturity date of the CDS, unless a credit event triggers termination of the CDS and a contingent payment from the protection seller to the protection buyer. If such a triggering event occurs, the protection buyer only pays a remaining fee for accrued protection from the last premium payment up to the time of the credit event. The settlement of the CDS then follows a pre-specified procedure, which was discussed in Chapter 2.

6.1 THEORETICAL PRICING APPROACH

A default swap, like an interest-rate swap, consists of two legs, one corresponding to the premium payments and the other to the contingent default payment. The present value (PV) of a default swap can be viewed as the algebraic sum of the present values of its two legs. The market premium is similar to an interest-rate swap in that the premium makes the current aggregate PV equal to zero. That is, for a par interest-rate swap, the theoretical net present value of the two legs must equal zero; the same principle applies for the two cash flow legs of a CDS.

The cash flows of a CDS are shown in Figure 6.1.

Normally, the default payment on a credit default swap will be $(1-\delta)$ times its notional amount, where δ is defined as the recovery rate of the reference security. The reason for this payout is clear — it allows a risky asset to be transformed into a risk-free asset by purchasing default protection referenced to this credit. For example, if the expected recovery rate for a given reference asset is 30% of its face value, upon default the

(A) *No default*

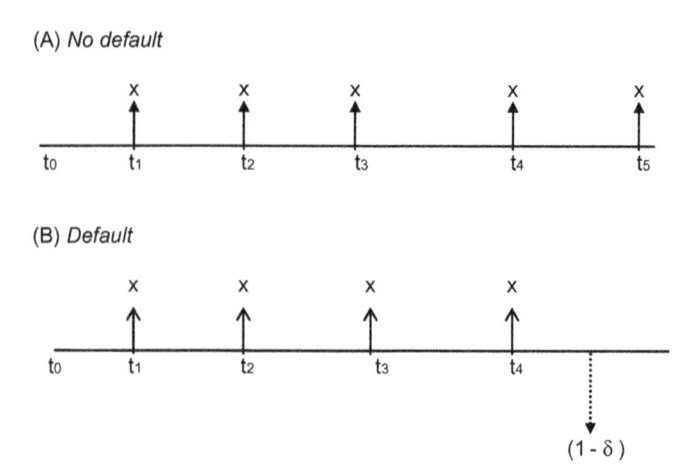

(B) *Default*

$(1 - \delta)$

FIGURE 6.1 Illustration of cash flows in a default swap.

remaining 70% will be paid by the protection seller. Credit agencies such as Moody's provide recovery rate estimates for corporate bonds with different credit ratings using historical data.

The valuation of each leg of the cash flow is considered below. As these cash flows may terminate at an unknown time during the life of the deal, their values are computed in a probabilistic sense, using the discounted expected value as calculated under the risk-neutral method and assumptions.

Here we introduce a reduced form type pricing model developed by Hull and White (2000). Their approach was to calibrate their model based on the traded bonds of the underlying reference name on a time series of credit default swap prices.

Like most other approaches, their model assumes that there is no counterparty default risk. Default probabilities, interest rates and recovery rates are independent.

Finally, they also assume that the claim in the event of default is the face value plus accrued interest. Consider the valuation of a plain vanilla credit default swap with a $1 notional principal. The notation used is:

T life of credit default swap in years
$q(t)$ risk-neutral probability density at time t
R expected recovery rate on the reference obligation in a risk-neutral world (independent of the time of default)
$u(t)$ present value of payments at the rate of $1 per year on payment dates between time zero and time t
$e(t)$ present value of an accrual payment at time t equal to $t - t^*$ where t^* is the payment date immediately preceding time t

6.1 THEORETICAL PRICING APPROACH

- $v(t)$ present value of $1 received at time t
- w total payment per year made by credit default swap buyer
- s value of w that causes the value of credit default swap to have a value of zero
- π the risk-neutral probability of no credit event during the life of the swap
- $A(t)$ accrued interest on the reference obligation at time t as a percentage of face value.

The value π is 1 minus the probability that a credit event will occur by time T. This is also referred to as the survival probability, and can be calculated from $q(t)$:

$$\pi = 1 - \int_0^a q(t)dt \tag{6.1}$$

The payments last until a credit event or until time T, whichever is sooner. If default occurs at t ($t < T$), the present value of the payment is $w[u(t) + e(t)]$. If there is no default prior to time T, the present value of the payment is $wu(T)$. The expected present value of the payment is therefore:

$$w \int_0^T q(t)[u(t) + e(t)]dt + w\pi u(T) \tag{6.2}$$

Given the assumption about the claim amount, the risk-neutral expected payoff from the CDS contract is derived as follows:

$$1 - R[1 + A(t)] \text{ multiplying } -R \text{ by } [1 + A(t)]$$

$$1 - R[1 + A(t)] = 1 - R - A(t)R$$

The present value of the expected payoff from the CDS is given as:

$$\int_0^T [1 - R - A(t)R]q(t)v(t)dt \tag{6.3}$$

The value of the credit default swap to the buyer is the present value of the expected payoff minus the present value of the payments made by the buyer, or:

$$\int_0^T [1 - R - A(t)R]q(t)v(t)dt - w \int_0^T q(t)[u(t) + e(t)]dt + w\pi u(T) \tag{6.4}$$

In equilibrium, the present value of each leg of the above equation should be equal. We can now calculate the credit default swap spread s, which is the value of w that makes the equation equal to zero, by simply rearranging the equation, as shown below.

$$s = \frac{\int_0^T [1 - R - A(t)R]q(t)v(t)dt}{\int_0^T q(t)[u(t) + e(t)]dt + \pi u(T)} \quad (6.5)$$

The variable s is referred to as the credit default swap spread, or CDS spread.

The formula in equation (6.5) is simple and intuitive for developing an analytical approach for pricing credit default swaps because of the assumptions used. For example, the model assumes that interest rates and default events are independent; also, the possibility of counterparty default is ignored. The spread s is the payment per year, as a percentage of the notional principal, for a newly issued credit default swap.

6.2 MARKET PRICING APPROACH[1]

We now present a discrete form pricing approach that is commonly used in the market, using market-observed parameter inputs, and is based on Professor Darrell Duffie's no-arbitrage approach first presented in 1992.

We stated earlier that a CDS has two cash-flow legs: the fee premium leg and the contingent cash-flow leg. We wish to determine the par spread or premium of the CDS, remembering that for a par spread valuation, in accordance with no-arbitrage principles, the net present value of both legs must be equal to zero (that is, they have the same valuation).

The valuation of the fee leg is given by the following relationship:

$$\text{PV of no-default fee payments} = s_N \times \text{Annuity}_N$$

which is given by:

$$PV = s_N \sum_{i=1}^{N} DF_i.PND_i.A_i \quad (6.6)$$

---

[1] A more descriptive explanation of this approach, with all steps described and including Excel spreadsheet formulae, is given in Appendix 6.1.

6.2 MARKET PRICING APPROACH

where:

- s_N is the par spread (CDS premium) for maturity N
- DF_i is the risk-free discount factor from time T_0 to time T_i
- PND_i is the no-default probability from T_0 to T_i
- A_i is the accrual period from T_{i-1} to T_i.

Note that the value for PND is for the specific reference entity for which a CDS is being priced.

If the accrual fee for the CDS is paid upon default and termination,[2] then the valuation of the fee leg is given by the relationship:

PV of no-default fee payments + PV of default accruals
$= S_N \times Annuity_N + S_N \times DefaultAccrual_N$

which is given by:

$$PV_{NoDefault+DefaultAccrual} = s_N \sum_{i=1}^{N} DF_i \cdot PND_i \cdot A_i \\ + s_N \sum_{i=1}^{N} DF_i \cdot (PND_{i-1} - PND_i) \cdot \frac{A_i}{2} \quad (6.7)$$

where:

$(PND_{i-1} - PND_i)$ is the probability of a credit event occurring during the period T_{i-1} to T_i

$\frac{A_i}{\{2\}}$ is the average accrual amount from T_{i-1} to T_i.

The valuation of the contingent leg is approximated by:

PV of Contingent $= Contingent_N$

which is given by:

$$PV_{Contingent} = (1 - R) \sum_{i=1}^{N} DF_i \cdot (PND_{i-1} - PND_i) \quad (6.8)$$

where R is the recovery rate of the reference obligation.

For a par credit default swap, we know that:

Valuation of fee leg = Valuation of contingent leg

---

[2]This is the amount of premium payable from the last payment date up to termination date, and similar to accrued coupon on a cash bond. Upon occurrence of a credit event and termination, the accrued premium to date is payable immediately. No protection payment is due from the protection seller until and after the accrual payment is made.

and therefore we can set:

$$s_N \sum_{i=1}^{N} DF_i \cdot PND_i \cdot A_i + s_N \sum_{i=1}^{N} DF_i \cdot (PND_{i-1} - PND_i) \cdot \frac{A_i}{2}$$
$$= (1-R) \sum_{i=1}^{N} DF_i \cdot (PND_{i-1} - PND_i) \quad (6.9)$$

which may be rearranged to give us the formula for the CDS premium s as follows:

$$s_N = (1-R) \sum_{i=1}^{N} \frac{DF_i \cdot (PND_{i-1} - PND_i)}{\sum_{i=1}^{N} DF_i \cdot PND_i \cdot A_i + DF_i \cdot (PND_{i-1} - PND_i) \cdot \frac{A_i}{2}} \quad (6.10)$$

In Table 6.1 we illustrate an application of the expression in equation (6.10) for a CDS of varying maturities, assuming a recovery rate of the defaulted reference asset of 30% and a given term structure of interest rates. It uses actual /360-day count convention.

For readers' reference we present a fuller explanation of this valuation approach in Appendix 6.1.

6.3 CREDIT DERIVATIVES PRICING IN VOLATILE ENVIRONMENTS: 'UPFRONT + CONSTANT SPREAD'

The foregoing and Appendix 6.1 show the standard approach to CDS pricing, and this conventional format is the 'all-running' format. This reflects that the CDS price is paid by the protection buyer through the life of the contract, as an annuity in effect, and there is no upfront payment at trade inception. A rough rule-of-thumb valuation at any time is given by:

Value = (Current spread − Initial spread) × Duration

The current and initial spreads are quoted data, but the duration must be calculated, and agreed between both counterparties. The calculation of duration is a function of the yield curve, the credit risky curve and the recovery rate. At low spreads the impact of these inputs on duration is negligible; at higher prices they can have a significant influence on duration, and hence the contract's value.

Following the 2007 liquidity crunch and the collapse of Lehman Brothers in 2008, CDS premiums reached very high levels, above 2000 bps in some cases. When a price spread rises above about 1200 bps, market practitioners do not value CDS on the basis of the 'all-running' approach, but instead use what is termed the

6.3 CREDIT DERIVATIVES PRICING

TABLE 6.1 Example of CDS Spread Premium Pricing

Maturity ts	Spot Rates	Discount Factors DFj	Survival Probability PSj	Default Probability	PV of Receipts if No Default	PV of Receipts if Default	Default Payment if Default	CDS Premium s
0.5	3.57%	0.9826	0.9993	0.0007	0.4910	0.0002	0.0005	0.10%
1.0	3.70%	0.9643	0.9983	0.0017	0.9723	0.0006	0.0016	0.17%
1.5	3.81%	0.9455	0.9972	0.0028	1.4437	0.0012	0.0035	0.24%
2.0	3.95%	0.9254	0.9957	0.0043	1.9044	0.0022	0.0063	0.33%
2.5	4.06%	0.9053	0.9943	0.0057	2.3545	0.0035	0.0099	0.42%
3.0	4.16%	0.8849	0.9932	0.0068	2.7939	0.0050	0.0141	0.50%
3.5	4.24%	0.8647	0.9900	0.0100	3.2220	0.0072	0.0201	0.62%
4.0	4.33%	0.8440	0.9886	0.0114	3.6392	0.0096	0.0269	0.74%
4.5	4.42%	0.8231	0.9859	0.0141	4.0450	0.0125	0.0350	0.86%
5.0	4.45%	0.8044	0.9844	0.0156	4.4409	0.0156	0.0438	0.98%
0.5	4.16%	0.9826	0.9993	0.0007	0.4910	0.0002	0.0005	0.10%
1.0	4.24%	0.9643	0.9983	0.0017	0.9723	0.0006	0.0016	0.17%

Recovery rate 0.3

108   6. CREDIT DEFAULT SWAP PRICING

'upfront + constant spread' format. This means that the price spread of the contract is constant and fixed by market convention (generally at 500 bps), and a (variable) upfront payment is made at trade inception. Under this approach, a CDS contract is quoted as its upfront value only (the constant spread, being constant, does not need to be quoted) and the valuation of the contract is no longer dependent on duration. The new approach is given by

Value = Quoted upfront − (Current spread − Initial spread) × Duration

This approach is also used for the iTraxx and CD-X indices and for equity tranches of structured products.

6.4 QUICK CDS CALCULATOR

Bloomberg has introduced a CDS calculator that does not reference a specific reference name; rather the purpose of the model is to calculate the cash flows or valuation associated with a CDS contract. The screen is QCDS, and it uses the International Swaps and Derivatives Association (ISDA) standard upfront settlement model and the

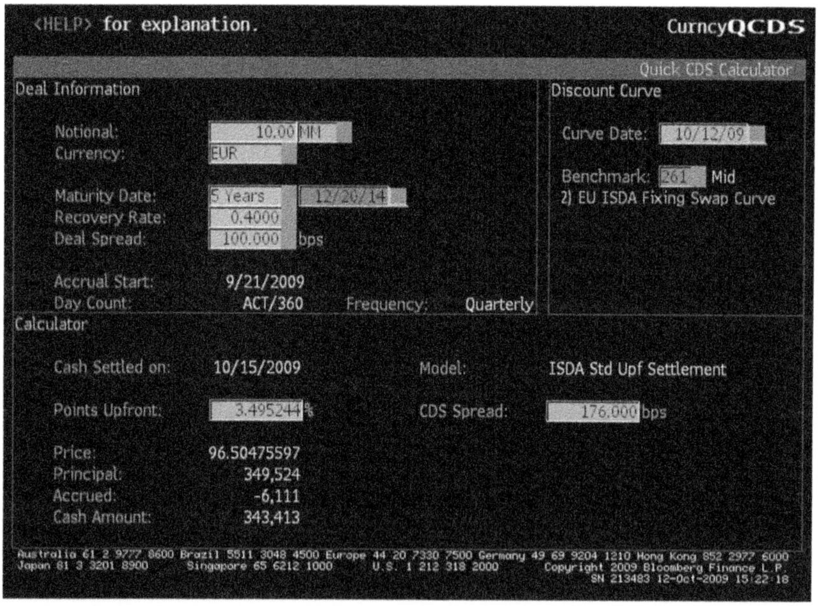

FIGURE 6.2 Bloomberg screen QCDS used to calculate points upfront, 12 October 2009. © *Bloomberg L.P. Reproduced with permission. Visit www.bloomberg.com*

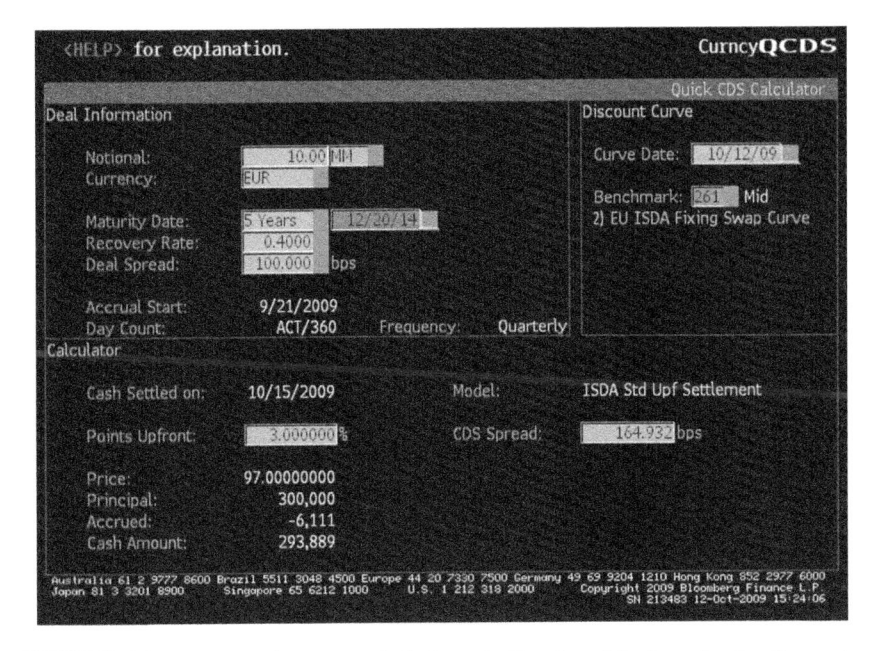

FIGURE 6.3 Change of inputs: calculation. © Bloomberg L.P. Reproduced with permission. Visit www.bloomberg.com

associated inputs to calculate the cash settlement amount and either the points upfront or the CDS spread.

The screen is shown at Figure 6.2. The user selects the currency, the relevant swap curve, the recovery rate and the deal spread (the fixed running coupon). The example we have shown is for a EUR-denominated CDS, hence defaults to the EUR swap curve, and is for 5 years. The user then inputs either the points upfront or the CDS spread, and the screen then calculates the other variable. In Figure 6.2 the 'deal spread' is 100 basis points and the CDS spread selected was 176 bps. The calculation indicates the equivalency between the points upfront and the CDS market spread. Figure 6.3 shows that when we input the points upfront at 3.00%, the CDS spread calculated was 164 bps.[4]

References and Bibliography

Choudhry, M. (2001). *The bond and money markets: Strategy, trading, analysis*. Oxford: Butterworth-Heinemann Elsevier.

[4]Coincidentally, the mid-price CDS spread for Allied Irish Bank 5-year CDS (in EUR) as at 12 October 2009.

Choudhry, M. (2003). Some issues in the asset-swap pricing of credit default swaps. In F. Fabozzi (Ed.), *Professional perspectives on fixed income portfolio management* (vol. 4). Hoboken, NJ: John Wiley & Sons.
Duffle, D. (1999). Credit swap valuation. *Financial Analysts Journal*, 73–87.
Duffle, D., & Huang, M. (1996). Swap rates and credit quality. *Journal of Finance*, 51, 3.
Hogg, R., & Craig, A. (1970). *Introduction to mathematical statistics* (3rd ed.). New York: Macmillan.
Hull, J., & White, A. (2000). Valuing credit default swaps I: No counter-party default risk. *Journal of Derivatives*, 8(1), 1–13.
Jarrow, R. A., & Turnbull, S. M. (1995). Pricing options on derivative securities subject to credit risk. *Journal of Finance*, 50, 53–58.
Longstaff, F. A., & Schwartz, E. S. (1995). Valuing credit derivatives. *Journal of Fixed Income*, 5(1), 6–12.
Yekutieli, I. (1999). With bond stripping, the curve's the thing. *Internal Bloomberg Report*.

APPENDICES

APPENDIX 6.1 THE MARKET APPROACH TO CDS PRICING

The market approach to CDS pricing adopts the same no-arbitrage concept as used in interest-rate swap pricing. This states that, at inception:

$$\text{PV Fixed leg} = \text{PV Floating leg}$$

Therefore for a CDS we set:

$$\text{PV Premium leg} = \text{PV Contingent leg}$$

The PV of the premium leg is straightforward to calculate, especially if there is no credit event during the life of the CDS. However, the contingent leg is just that — contingent on occurrence of a credit event. Hence we need to determine the value of the premium leg at time of the credit event. This requires us to use default probabilities. We can use historical default rates to determine default probabilities, or back them out using market CDS prices. The latter approach is in fact *implied probabilities*.

Default probabilities

To price a CDS, we need the answers to two basic questions:

- What is the probability of a credit event?
- If a credit event occurs, how much is the protection seller likely to pay?
  This revolves around an assumed *recovery rate*.
  We may also need to know:
- If a credit event occurs, when does this happen?

Let us consider first the probability of default. One way to obtain default probabilities is to observe credit spreads in the corporate bond market. Riskless investments establish a benchmark riskless interest rate, usually the government bond yield. In the corporate (non-zero default probability) lenders and investors expect to receive a higher return from risky investments. The difference between the risky and risk-less rates is the *credit spread*. The credit spread will vary according to:

- credit quality (e.g. credit rating);
- maturity;
- liquidity;
- supply and demand.

Of these factors, one of the most significant is the term to maturity. The *term structure of credit spreads* exhibits a number of features. For instance, lower-quality credits trade at a wider spread than higher-quality credits, and longer-dated obligations normally have higher spreads than shorter-dated ones. For example, for a particular sector they may look like this:

- 2yr AA: 20 bp;
- 5yr AA: 30 bp;
- 10yr AA: 37 bp.

An exception to this is at the very low end of the credit spectrum, for example we may observe the following yields for CCC-rated assets:

- 2yr CCC: 11%;
- 5yr CCC: 7.75%;
- 10yr CCC: 7%.

In the case of the CCC rating this reflects the belief that there is a higher probability of default risk right now rather than 5 years from now, because if the company survives the first few years, the risk of later default is much lower later on. This gives rise to lower spreads.

Suppose that the corporate bonds of a particular issuer trade at the yields shown in Table 6A.1.

We calculate the continuously compounded rate of return on the risk-free asset to be:

$$e^{rt}$$

The rate of return on the risky asset is therefore given by:

$$e^{(r+y)t}$$

We now calculate the default probability assuming zero recovery of the asset value following default. On this assumption, if the probability of

6. CREDIT DEFAULT SWAP PRICING

TABLE 6A.1 Hypothetical Corporate Bond Yields and Risk Spread

Maturity t	Risk-Free Yield r	Corporate Bond Yield $r+y$	Risk Spread y
0.5	3.57%	3.67%	0.10%
1.0	3.70%	3.82%	0.12%
1.5	3.81%	3.94%	0.13%
2.0	3.95%	4.10%	0.15%
2.5	4.06%	4.22%	0.16%
3.0	4.16%	4.32%	0.16%
3.5	4.24%	4.44%	0.20%
4.0	4.33%	4.53%	0.20%
4.5	4.42%	4.64%	0.22%
5.0	4.45%	4.67%	0.22%

default is p, then an investor should be indifferent between an expected return of:

$$(1-p)e^{(r+y)t}$$

on the risky corporate bond, and:

$$e^{rt}.$$

Setting these two expressions equal we have:

$$(1-p)e^{(r+y)t} = e^{rt} \qquad (A.1)$$

Solving for p gives:

$$p = 1 - e^{-yt} \qquad (A.2)$$

Using $p = 1 - e^{-yt}$ we can calculate therefore the probabilities of default from the credit spreads that were shown in Table 6A.1. These are shown in Table 6A.2.

For example,

$$p_{0,5} = 1 - e^{-0.0025 \times 5} = 1.094\%$$

is the cumulative probability of default over the complete 5-year period, while:

$$p_{4,5} = p_{0,5} - p_{0,4} = 0.109\%$$

is the probability of default in year 5.

We then extend the analysis to an assumption of a specified recovery rate following default. If the probability of default is p, and the recovery rate is R, then an investor should now be indifferent between an expected return of:

$$(1-p)e^{(r+y)t} + Rpe^{(r+y)t} \quad (A.3)$$

on the risky corporate bond, and e^{rt} on the (risk-free) government bond.

Again, setting these two expressions equal, and solving for p gives:[5]

$$(1-p)e^{(r+y)t} + Rpe^{(r+y)t} = e^{rt}$$
$$p = \frac{1-e^{-yt}}{1-R} \quad (A.4)$$

Using this formula and assuming a recovery rate of 30% we calculate the cumulative default probabilities shown in Table 6A.3.

For example:

$$p_{0,5} = \frac{1-e^{-0.0022 \times 5}}{1-0.30} = 1.563\%$$

is the cumulative probability of default over the 5-year period.

We now expand the analysis to default and survival probabilities. Consider what happens to a risky asset over a specific period of time: there are just two possibilities, which are:

- There is a credit event, and the asset defaults.
- There is no credit event, and the asset survives.

---

[5]The steps in between are:
$$(1-p)e^{(r+y)t} = e^{rt}$$
$$(1-p)e^{rt} \cdot e^{yt} = e^{rt}$$
$$(1-p)e^{-yt} = 1$$
$$1-p = e^{-yt}$$
$$p = 1 - e^{-yt}$$

Incorporating the recovery rate R we have the following steps:
$$1 - p + pR = e^{-yt}$$
$$-p + pR = e^{-yt} - 1$$
$$-p(1-R) = e^{-yt} - 1$$
$$-p = \frac{e^{-yt} - 1}{1-R}$$
$$p = \frac{1-e^{-yt}}{1-R}$$

TABLE 6A.2 Default Probabilities

Maturity t	Risk-Free Yield r	Corporate Bond Yield r+y	Risk Spread y	Cumlative Probability of Default	Annual Probability of Default
0.5	3.57%	3.67%	0.10%	0.050%	0.050%
1.0	3.70%	3.82%	0.12%	0.120%	0.070%
1.5	3.81%	3.94%	0.13%	0.195%	0.075%
2.0	3.95%	4.10%	0.15%	0.299%	0.104%
2.5	4.06%	4.22%	0.16%	0.399%	0.100%
3.0	4.16%	4.32%	0.16%	0.479%	0.080%
3.5	4.24%	4.44%	0.20%	0.698%	0.219%
4.0	4.33%	4.53%	0.20%	0.797%	0.099%
4.5	4.42%	4.64%	0.22%	0.985%	0.188%
5.0	4.45%	4.67%	0.22%	1.094%	0.109%

TABLE 6A.3 Cumulative Default Probabilities

Maturity t	Risk-Free Yield r	Corporate Bond Yield r+y	Risky Spread	Cumulative Probability of Default
0.5	3.57%	3.67%	0.10%	0.071%
1.0	3.7%	3.82%	0.12%	0.171%
1.5	3.81%	3.94%	0.13%	0.279%
2.0	3.95%	4.10%	0.15%	0.427%
2.5	4.06%	4.22%	0.16%	0.570%
3.0	4.16%	4.32%	0.16%	0.684%
3.5	4.24%	4.44%	0.20%	0.997%
4.0	4.33%	4.53%	0.20%	1.139%
4.5	4.42%	4.64%	0.22%	1.407%
5.0	4.45%	4.67%	0.22%	1.563%

Let us call these outcomes D (for default), having a probability q, and S (for survival) having probability of $(1 - q)$. We can represent this as a binary process, shown as Figure 6A.1.

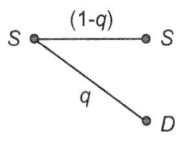

FIGURE 6A.1 Binary process of survival or default.

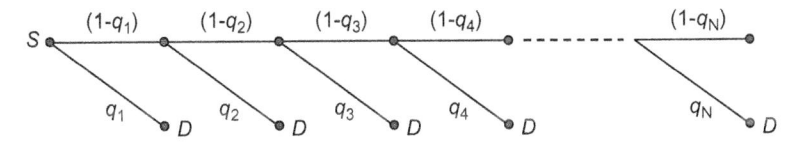

FIGURE 6A.2 Binary process of survival or default over multiple periods.

Over multiple periods this binary process can be illustrated as shown in Figure 6A.2.

As shown in Figure 6A.2, the probability of survival to period N is then:

$$PSN = (1 - q1) \times (1 - q2) \times (1 - q3) \times (1 - q4) \times \ldots \times (1 - qN) \quad (A.5)$$

while the probability of default in any period N is:

$$PSN - 1 \times qN = PSN - 1 - PSN \quad (A.6)$$

Given these formulas, we can now price a CDS contract.

Pricing CDS contract

Given a set of default probabilities, we can calculate the fair premium for a CDS, which is the market approach first described in Chapter 1. To do this, consider a CDS as a series of contingent cash flows, the cash flows depending upon whether a credit event occurs. This is shown as Figure 6A.3. The symbols are:

s   is the CDS premium
k   is the day count fraction when default occurred
R   is the recovery rate

We wish first to value the premium stream given no default, shown at Figure 6A.3(a). The expected PV of the stream of CDS premiums over time can be calculated as:

$$PVS_{nd} = s \sum_{j=1}^{N} DF_j PS_j T_{j-1,j} \quad (A.7)$$

(A) No default

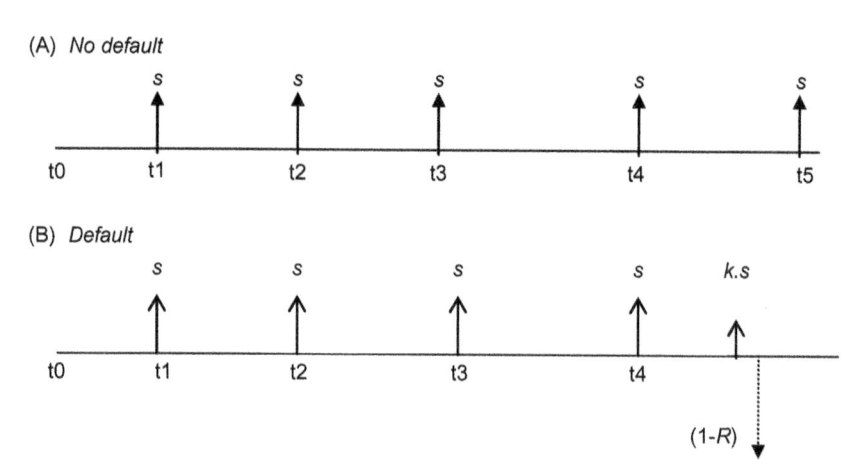

```
         s       s       s              s              s
         ↑       ↑       ↑              ↑              ↑
─────────┼───────┼───────┼──────────────┼──────────────┼──────
   t0    t1      t2      t3             t4             t5
```

(B) Default

```
         s       s       s              s      k.s
         ↑       ↑       ↑              ↑       ↑
─────────┼───────┼───────┼──────────────┼───────┼──────────
   t0    t1      t2      t3             t4
                                                 │
                                           (1-R) ↓
```

FIGURE 6A.3 CDS contingent cash flows.

where:

PVS_{nd} is the expected present value of the stream of CDS premiums if there is no default

s is the CDS spread (fee or premium)

DF_j is the discount factor for period j

PS_j is the probability of survival through period j

$T_{j-1,j}$ is the length of time of period j (expressed as a fraction of a year)

We now require an expression for the value of the premium stream given default, which are the cash flows shown at Figure 6A.3(b). If a default occurs half-way through period C, and the CDS makes the default payment at the end of that period, the expected present-value of the fees received is:

$$PVS_d = s \sum_{j=1}^{C} DF_j PS_j T_{j-1,j} + s \cdot DF_C PD_C \frac{T_{C-1,C}}{2} \quad (A.8)$$

while the value of the default payment is:

$$(1 - R) DF_C PD_C \quad (A.9)$$

where:

PVS_d is the expected present-value of the stream of CDS premiums if there is default in period C

PD_C is the probability of default in period C

R is the recovery rate

and the other terms are as before.

On the no-arbitrage principle, which is the same approach used to price interest-rate swaps, for a CDS to be fairly priced the expected value of the premium stream must equal the expected value of the default payment.

As default can occur in any period j, therefore we can write:

$$s \sum_{j=1}^{N} DF_j PS_j T_{j-1,j} + s \sum_{j=1}^{N} DF_j PD_j \frac{T_{j-1,j}}{2} = (1-R) \sum_{j=1}^{N} DF_j PD_j \quad (A.10)$$

In equation (A.10) the first part of the left-hand side (LHS) is the expected present-value of the stream of premium payments if no default occurs, and the second part of the LHS is the expected present-value of the accrued premium payment in the period when default occurs. The right-hand side of (A.10) is the expected present-value of the default payment in the period when default occurs.

Rearranging this expression gives the fair premium s for the CDS shown as (A.11):

$$s = \frac{(1-R) \sum_{j=1}^{N} DF_j PD_j}{\sum_{j=1}^{N} DF_j PS_j T_{j-1,j} + \sum_{j=1}^{N} DF_j PD_j \frac{T_{j-1,j}}{2}} \quad (A.11)$$

Example calculation

We have shown then that the price of a CDS contract can be calculated from the spot rates and default probability values given earlier. In this example we assume that the credit event (default) occurs half-way through the premium period, thus enabling us to illustrate the calculation of the present-value of the receipt in event of default (the second part of the left-hand side of the original no-arbitrage equation (A.10), the accrual factor) in more straightforward fashion.

Table 6A.4 illustrates the pricing of a CDS contract written on the reference entity whose credit spread premium over the risk-free rate was introduced earlier. The default probabilities were calculated as shown in Table 6A.3.

Table 6A.5 shows the Microsoft Excel formulae used in the calculation spreadsheet.

Consider the 1-year CDS premium. From Table AII.6 1-year CDS premium is 0.17%. To check this calculation, we observe the expected present-value of the premium for the 6-month and 1-year dates, which is:

- Survival probability × Discount factor × Premium × Day count fraction
  For the 6-month period this is 0.9993 × 0.9826 × 0.0017 × 0.5 or 0.0008346.
  For the 1-year period this is 0.9983 × 0.9643 × 0.1017 × 0.5 or 0.00081826.

6. CREDIT DEFAULT SWAP PRICING

TABLE 6A.4 Calculation of CDS Prices

	A	B	C	D	E	F	G	H	I
							\multicolumn{2}{c	}{Probability-Weighted PVs}	
	Maturity t	Spot Rates	Discount Factors DF_j	Survival Probability PS_j	Default Probability PD_j	PV of Receipts if No Default	PV of Receipt if Default	Default Payment if Default	CDS Premiums
1	0.5	3.57%	0.9826	0.9993	0.0007	0.4910	0.0002	0.0005	0.10%
2	1.0	3.70%	0.9643	0.9983	0.0017	0.9723	0.0006	0.0016	0.17%
3	1.5	3.81%	0.9455	0.9972	0.0028	1.4437	0.0012	0.0035	0.24%
4	2.0	3.95%	0.9254	0.9957	0.0043	1.9044	0.0022	0.0063	0.33%
5	2.5	4.06%	0.9053	0.9943	0.0057	2.3545	0.0035	0.0099	0.42%
6	3.0	4.16%	0.8849	0.9932	0.0068	2.7939	0.0050	0.0141	0.50%
7	3.5	4.24%	0.8647	0.9900	0.0100	3.2220	0.0072	0.0201	0.62%
8	4.0	4.33%	0.8440	0.9886	0.0114	3.6392	0.0096	0.0269	0.74%
9	4.5	4.42%	0.8231	0.9859	0.0141	4.0450	0.0125	0.0350	0.86%
10	5.0	4.45%	0.8044	0.9844	0.0156	4.4409	0.0156	0.0438	0.98%

Recovery rate 0.3

AN INTRODUCTION TO CREDIT DERIVATIVES

TABLE 6A.5 CDS Price Calculation: Excel Spreadsheet Formulae

	A	B	C	D	E	F	G	H	I
							Probability-Weighted PVs		
	Maturity t	Spot Rates	Discount Factors DF_j	Survival Probability PS_j	Default Probability PD_j	PV of Receipts if No Default	PV of Receipt if Default	Default Payment if Default	CDS Premium s
1	0.5	3.57%	=1/(1+C6)^0.5	=1-F6	0.0007	=SUMPRODUCT($D$6:D6,$E$6:E6)*0.5	=SUMPRODUCT($D$6:D6,$F$6:F6)*0.5/2	=(1-$B$18)*SUMPRODUCT($D$6:D6,$F$6:F6)	=I6/(G6+H6)
2	1.0	3.70%	=1/(1+C7)^1	=1-F7	0.0017	=SUMPRODUCT($D$6:D7,$E$6:E7)*0.5	=SUMPRODUCT($D$6:D7,$F$6:F7)*0.5/2	=(1-$B$18)*SUMPRODUCT($D$6:D7,$F$6:F7)	=I7/(G7+H7)
3	1.5	3.81%	=1/(1+C8)^1.5	=1-F8	0.0028	=SUMPRODUCT($D$6:D8,$E$6:E8)*0.5	=SUMPRODUCT($D$6:D8,$F$6:F8)*0.5/2	=(1-$B$18)*SUMPRODUCT($D$6:D8,$F$6:F8)	=I8/(G8+H8)
4	2.0	3.95%	=1/(1+C9)^2	=1-F9	0.0043	=SUMPRODUCT($D$6:D9,$E$6:E9)*0.5	=SUMPRODUCT($D$6:D9,$F$6:F9)*0.5/2	=(1-$B$18)*SUMPRODUCT($D$6:D9,$F$6:F9)	=I9/(G9+H9)
5	2.5	4.06%	=1/(1+C10)^2.5	=1-F10	0.0057	=SUMPRODUCT($D$6:D10,$E$6:E10)*0.5	=SUMPRODUCT($D$6:D10,$F$6:F10)*0.5/2	=(1-$B$18)*SUMPRODUCT($D$6:D10,$F$6:F10)	=I10/(G10+H10)
6	3.0	4.16%	=1/(1+C11)^3	=1-F11	0.0068	=SUMPRODUCT($D$6:D11,$E$6:E11)*0.5	=SUMPRODUCT($D$6:D11,$F$6:F11)*0.5/2	=(1-$B$18)*SUMPRODUCT($D$6:D11,$F$6:F11)	=I11/(G11+H11)

(Continued)

TABLE 6A.5 CDS Price Calculation: Excel Spreadsheet Formulae (Continued)

	A	B	C	D	E	F	G	H	I
							Probability-Weighted PVs		
	Maturity t	Spot Rates	Discount Factors DF_j	Survival Probability PS_j	Default Probability PD_j	PV of Receipts if No Default	PV of Receipt if Default	Default Payment if Default	CDS Premium s
7	3.5	4.24%	=1/(1+C12)^3.5	=1-F12	0.0100	=SUMPRODUCT($D$6:D12,$E$6:E12)*0.5	=SUMPRODUCT($D$6:D12,$F$6:F12)*0.5/2	=(1-$B$18)*SUMPRODUCT($D$6:D12,$F$6:F12)	=I12/(G12+H12)
8	4.0	4.33%	=1/(1+C13)^4	=1-F13	0.0114	=SUMPRODUCT($D$6:D13,$E$6:E13)*0.5	=SUMPRODUCT($D$6:D13,$F$6:F13)*0.5/2	=(1-$B$18)*SUMPRODUCT($D$6:D13,$F$6:F13)	=I13/(G13+H13)
9	4.5	4.42%	=1/(1+C14)^4.5	=1-F14	0.0141	=SUMPRODUCT($D$6:D14,$E$6:E14)*0.5	=SUMPRODUCT($D$6:D14,$F$6:F14)*0.5/2	=(1-$B$18)*SUMPRODUCT($D$6:D14,$F$6:F14)	=I14/(G14+H14)
10	5.0	4.45%	=1/(1+C15)^5	=1-F15	0.0156	=SUMPRODUCT($D$6:D15,$E$6:E15)*0.5	=SUMPRODUCT($D$6:D15,$F$6:F15)*0.5/2	=(1-$B$18)*SUMPRODUCT($D$6:D15,$F$6:F15)	=I15/(G15+H15)

Recovery rate 0.3

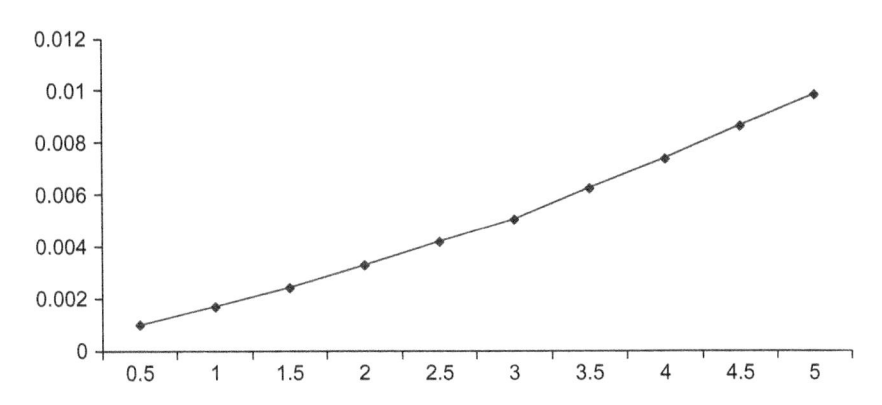

FIGURE 6A.4 Term structure of credit rates.

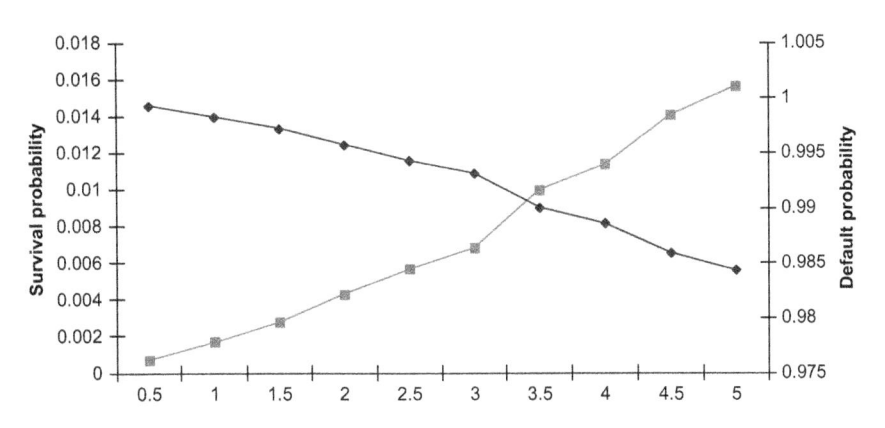

FIGURE 6A.5 Term structure of default probabilities.

The expected present-value of the accrued premium if default occurs half-way through a period is:
- Default probability × Discount factor × Premium × Day count fraction
For the 6-month period this is $0.0007 \times 0.9826 \times 0.0017 \times 0.25$ which actually comes out to a negligible value. For the 1-year period the amount is:

$0.0017 \times 0.9643 \times 0.0017 \times 0.25$ which is also negligible.

The total expected value of premium income is **0.00166.**

The expected present-value of the default payment if payment is made at end of the period is:
- Default probability × Discount factor × (1 − Recovery rate), which for the two periods is:
- 6mth: $0.0007 \times 0.9826 \times (1-30\%) = 0.000482$
- 12mth: $0.0017 \times 0.9643 \times (1-30\%) = 0.001148$

So the total expected value of the default payment is **0.00166**, which is equal to the earlier calculation. Our present values for both fixed leg and contingent legs are identical, which means we have the correct no-arbitrage value for the CDS contract.

From the CDS premium values we can construct a term structure of credit rates for this particular reference credit (or reference sector), which is shown in Figure 6A.4. We can also construct a term structure of default probabilities, and this shown in Figure 6A.5.

CHAPTER 7

The Asset Swap—Credit Default Swap Basis[1]

We saw in Chapter 2 that asset swaps, although pre-dating the credit derivative market, are a form of credit derivative but are in fact viewed as cash market instruments. However, because an asset swap is a structure that expllicitly prices a credit-risky bond in terms of its spread over Libor (inter-bank credit risk), in theory it can be viewed as a means by which to price credit derivatives. In fact, in the early days of the credit derivatives market the most common method of pricing credit default swaps (CDSs) was by recourse to the asset swap spread of the reference credit, as the CDS premium should (in theory) be equal to the asset swap spread of the reference asset. Certainly we can say that the asset swap provides an indicator of the minimum returns that would be required for specific reference credits, as well as a mark-to-market reference. It is also a hedging tool for a CDS position.

We first consider the use of this technique, before observing how these two spread levels known as the *CDS basis* differ in practice.

7.1 ASSET SWAP PRICING

7.1.1 Basic Concept

Credit derivatives were originally marked using the asset-swap pricing technique. The asset-swap market is a reasonably reliable indicator of the returns required for individual credit exposures, and provides a mark-to-market framework for reference assets as well as a hedging

[1] Parts of this chapter first appeared in Choudhry (2001).

mechanism. As we saw in Chapter 2, a par asset swap typically combines the sale of an asset such as a fixed-rate corporate bond to a counterparty, at par and with no interest accrued, with an interest-rate swap. The coupon on the bond is paid in return for Libor, plus a spread if necessary. This spread is the asset-swap spread, and is the price of the asset swap. In effect, the asset swap allows market participants that pay Libor-based funding to receive the asset-swap spread. This spread is a function of the credit risk of the underlying bond asset, which is why it may be viewed as equivalent to the price payable on a CDS written on that asset.

The generic pricing is given by:

$$Y_a = Y_b - ir \quad (7.1)$$

where:

Y_a is the asset swap spread
Y_b is the asset spread over the benchmark
ir is the interest-rate swap spread

The asset spread over the benchmark is simply the bond (asset) redemption yield over that of the government benchmark. The interest-rate swap spread reflects the cost involved in converting fixed-coupon benchmark bonds into a floating-rate coupon during the life of the asset (or default swap), and is based on the swap rate for that maturity.

The theoretical basis for deriving a default swap price from the asset swap rate can be illustrated by looking at a basis-type trade involving a cash market reference asset (bond) and a default swap written on this bond. This is similar in concept to the risk-neutral or *no-arbitrage* concept used in derivatives pricing. The theoretical trade involves:

- a long position in the cash market floating-rate note (FRN) priced at par, and which pays a coupon of Libor + X basis points;
- a long position (bought protection) in a credit default swap written on the same FRN, of identical term-to-maturity and at a cost of Y basis points.

The buyer of the bond is able to fund the position at Libor. In other words, the bondholder has the following net cash flow:

$$(100 - 100) + ((\text{Libor} + X) - (\text{Libor} + Y)) \quad (7.2)$$

or $X - Y$ basis points.

In the event of default, the bond is delivered to the protection seller in return for payment of par, enabling the bondholder to close out the funding position. During the term of the trade the bondholder has earned $X - Y$ basis points while assuming no credit risk. For the trade to meet the no-arbitrage condition, we must have $X = Y$. If $X \neq Y$, the investor would be able to establish the position and generate a risk-free profit.

This is a logically tenable argument as well as a reasonable assumption. The default risk of the cash bondholder is identical in theory to that of the default seller. In the next section, we illustrate an asset-swap pricing example before looking at why, in practice, there exist differences in pricing between default swaps and cash market reference assets.

7.1.2 Asset-Swap Pricing Example

XYZ plc is a Baa2-rated corporate. The 7-year asset swap for this entity is currently trading at 93 basis points; the underlying 7-year bond is hedged by an interest-rate swap with an Aa2-rated bank. The risk-free rate for floating-rate bonds is Libid minus 12.5 basis points (assume the bid-offer spread is 6 basis points). This suggests that the credit spread for XYZ plc is 111.5 basis points. The credit spread is the return required by an investor for holding the credit of XYZ plc. The protection seller is conceptually long on the asset, and so would short the asset to hedge its position. This is illustrated in Figure 7.1. The price charged for the default swap is the price of shorting the asset, which works out as 111.5 basis points each year.

Therefore we can price a credit default written on XYZ plc as the present value of 111.5 basis points for 7 years, discounted at the interest-rate swap rate of 5.875%. This computes to a credit default swap price of 6.25%:

Reference	XYZ plc
Term	7 years
Interest-rate swap rate	5.875%
Asset swap Libor plus	93 bps
CREDIT DEFAULT SWAP PRICING	
Benchmark rate	Libid minus 12.5 bps
Margin	6 bps
Credit default swap	111.5 bps
Default swap price	6.252%

7.1.3 Pricing Differentials

Market observation shows us that, contrary to what theory tells us, the prices of asset swaps and CDS contracts written on the same corporate entity differ, sometimes by a considerable extent. This should not

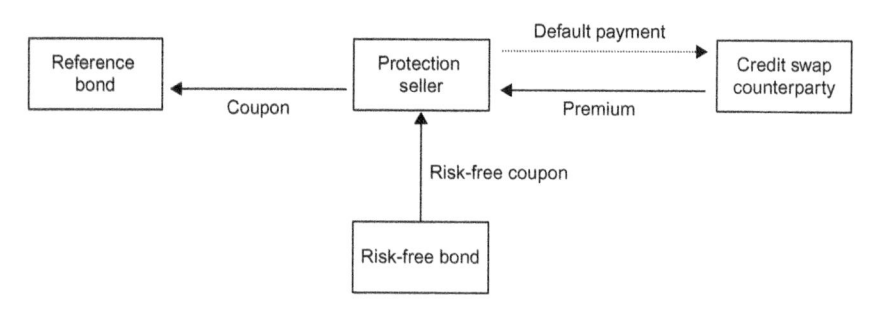

FIGURE 7.1 Credit default swap and asset swap hedge.

surprise us, once we stop to think about it. A number of factors observed in the market serve to make the price of credit risk that has been established synthetically using CDS differ from its price as traded in the cash market. In fact, identifying (or predicting) such differences gives rise to arbitrage opportunities that may be exploited by basis trading in the cash and derivative markets.[2] These factors include the following:

- Bond identity — the bondholder is aware of the exact issue that it is holding in the event of default, however credit default swap sellers may receive potentially any bond from a basket of deliverable instruments that rank *pari passu* with the cash asset; this is the delivery option afforded the long swap holder.
- The borrowing rate for a cash bond in the repo market may differ from Libor if the bond is to any extent *special;* this does not impact the credit default swap price, which is fixed at inception.
- Certain bonds rated AAA (such as World Bank securities) sometimes trade below Libor in the asset swap market; however, a bank writing protection on such a bond will expect a premium (positive spread over Libor) for selling protection on the bond.
- Depending on the precise reference credit, the credit default swap may be more liquid than the cash bond (resulting in a lower credit default swap price) or less liquid than the bond (resulting in a higher price).
- Credit default swaps may be required to pay out on credit events that are technical defaults, and not the full default that impacts a cash bondholder; protection sellers may demand a premium for this additional risk.
- The credit default swap buyer is exposed to counterparty risk during the term of the trade, unlike the cash bondholder.

[2]This is known as trading the credit default basis, and involves either buying the cash bond and buying a CDS written on this bond, or selling the cash bond and selling a default swap written on the bond. This is covered in Choudhry (2006).

- After the financial crash, the value of actual cash has risen, so asset swap (ASW) levels are usually higher than CDS levels on the same name but not always.

For these and other reasons, the default swap price always differs from the cash market price for the same asset. In any case, the existence of the basis means that banks generally use credit pricing models, based on the same models used to price interest-rate derivatives, when pricing credit derivatives.

7.2 THE BASIS AS MARKET INDICATOR

Contrary to the no-arbitrage logic that might be expected to prevail, the CDS-bond basis does not trade close to zero, and can exhibit quite large values. The factors that cause the basis to diverge away from a zero value are, as we saw earlier, a combination of contract-specific issues and market issues. Simple supply–demand factors and their impact in the cash and synthetic markets also influence the basis.

Market participants observe the basis closely, both as an indicator of current market levels and also as a predictor of future direction. This reflects the fact that the basis is an indicator of general market sentiment and its appetite for cash versus synthetic risk. The relationship between the two markets flows both ways, that is absolute values as well as the spread may widen first in the CDS market before moving onto the cash market, and *vice versa*. For this reason the basis sometimes underperforms cash when spreads start to widen (such as going into a period of recession) while at other times it may outperform it. Generally though, the basis is a leading indicator of the market rather than a lagging one, essentially because it is part of what is often a more liquid market compared to the cash bonds in the same sector. For instance, the 5-year benchmark CDS can be traded in large size without having an adverse impact on the market. In many cases the CDS is easier to source in the market compared to the same-name cash bond.

The indicative nature of the basis is shown in Figure 7.2. This shows the average basis in the investment-grade US industrials sector during 2002−2003, calculated from a set of 50 reference names, compared to an investment bank corporate bond index for the same asset class.

One method we can use to measure the pace of change in the basis at any time is to check the change in the basis itself against the option-adjusted spread (OAS) for the sector.[3] That is, we study the relationship between changes in the basis for a particular sector and the change in

---

[3] More information on OAS can be found in Choudhry (2004).

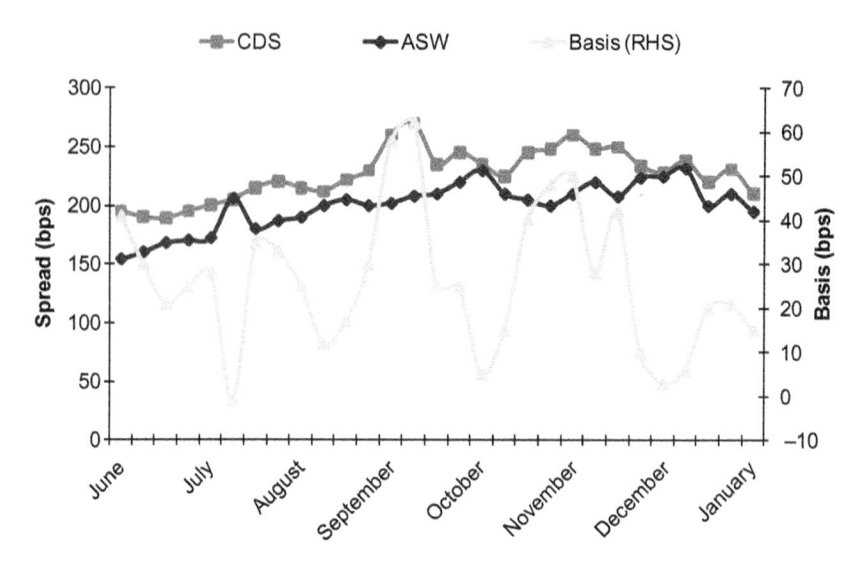

FIGURE 7.2 The dynamics of the credit default swap basis, US IG-rated industrials index, 2002.

that sector's OAS. Essentially, the liquidity of the synthetic market means that the CDS market is a leading indicator of market sentiment, compared to the cash market, and would appear to react more quickly to market sentiment. As such, portfolio managers observe the basis levels closely for an idea of future direction and sentiment. As the aggregate value of the basis has a low correlation with the direction of spreads, it can be considered sensible diversification for the portfolio manager. This would normally be in the form of a basis trade of the type described at the start of this chapter. As the value of the basis is only weakly influenced by the market direction for credit, basis trades can be considered efficient portfolio allocation for some fund managers.

Figure 7.3 shows a regression analysis of weekly changes in the basis compared to the weekly changes in the cash OAS for that sector. We see that the two variables are only very weakly correlated. Therefore, a portfolio manager may consider a basis-type trade as essential diversification, as well as useful for market intelligence purposes.

7.3 ANALYZING THE BASIS SPREAD MEASURE

A wide range of factors drive the basis, some of which were described earlier in the chapter.

The existence of a non-zero basis has implications for investment strategy. For instance, when the basis is negative investors may prefer

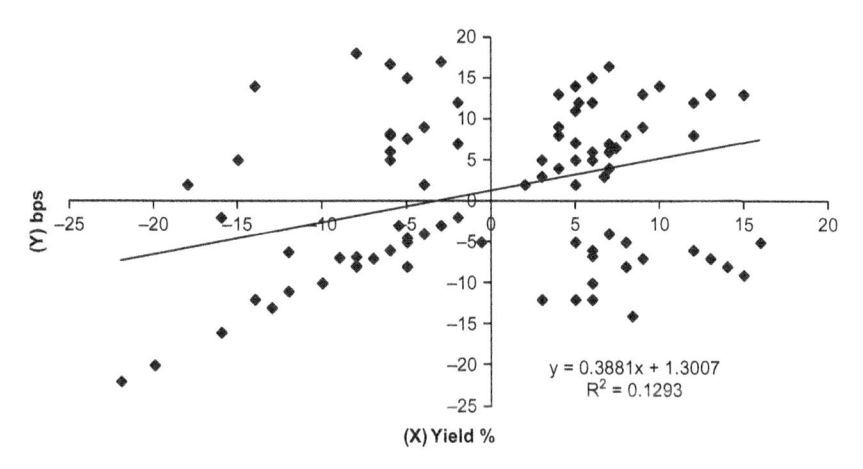

FIGURE 7.3 Regression analysis of weekly changes in the basis compared to weekly changes in the cash OAS.

to hold the cash bond, whereas if for liquidity, supply or other reasons the basis is positive the investor may wish to hold the asset synthetically, by selling protection using a credit default swap. One of the most common causal factors of a positive basis is a cash bond that is overvalued, perhaps due to excess demand over supply.

Another approach is to arbitrage between the cash and synthetic markets, in the case of a negative basis by buying the cash bond and shorting it synthetically by buying protection in the CDS market. Investors have a range of spreads to use when performing their relative value analysis. An accurate measure of the basis is vital for an effective assessment of the relationship between the two markets to be made.

7.3.1 ASW Spread

The traditional method used to calculate the basis given by equation (7.1) can be regarded as being of sufficient accuracy only if the following conditions are satisfied:

- the CDS and bond are of matching maturities;
- for many reference names, both instruments carry a similar level of subordination.

The conventional approach for analyzing an asset swap uses the bond's yield-to-maturity (YTM) in calculating the spread. The assumptions implicit in the YTM calculation also make this spread problematic for relative analysis.

The most critical issue however, is the nature of the construction of the asset swap structure itself: the need for the cash bond price to be priced at or very near to par. Most corporate bonds trade significantly away from par, thus rendering the par asset swap price an inaccurate measure of their credit risk. The standard asset swap is constructed as a par product; hence if the bond being asset-swapped is trading above par then the swap price will overestimate the level of credit risk. If the bond is trading below par the asset swap will underestimate the credit risk associated with the bond. We see then that using the CDS–ASW method will provide an unreliable measure of the basis. We therefore need to use another methodology for measuring the basis.

7.3.2 Z-Spread

A commonly used alternative to the ASW spread is the Z-spread. The Z-spread uses the zero-coupon yield curve to calculate spread, so in theory is a more effective spread to use. The zero-coupon curve used in the calculation is derived from the interest-rate swap curve. Put simply, the Z-spread is the basis point spread that would need to be added to the implied spot yield curve such that the discounted cash flows of the a bond are equal to its present value (its current market price). Each bond cash flow is discounted by the relevant spot rate for its maturity term. How does this differ from the conventional asset-swap spread? Essentially, in its use of zero-coupon rates when assigning a value to a bond. Each cash flow is discounted using its own particular zero-coupon rate. The price of a bond's price at any time can be taken to be the market's value of the bond's cash flows. Using the Z-spread we can quantify what the swap market thinks of this value, that is, by how much the conventional spread differs from the Z-spread. Both spreads can be viewed as the coupon of a swap market annuity of equivalent credit risk of the bond being valued.

In practice the Z-spread, especially for shorter-dated bonds and for higher credit-quality bonds, does not differ greatly from the conventional asset-swap spread. The Z-spread is usually the higher spread of the two, following the logic of spot rates, but not always. If it differs greatly, then the bond can be considered to be mis-priced.

7.3.3 Critique Of The Z-Spread

Although the Z-spread is an improved value to use when calculating the basis compared to the ASW spread, it is not a direct comparison with a CDS premium. This is because it does not allow for probability of default, or more specifically timing of default. The Z-spread uses the correct zero-coupon rates to discount each bond cash flow, but it does not reflect the fact that, as coupons are received over time, each cash flow will carry a different level of credit risk. In fact, each cash flow will represent different levels of credit risk.

Given that a corporate bond that pays a spread over Treasuries or Libor must, by definition, be assumed to carry credit risk, we need to incorporate a probability of default factor if we compare its value to that of the same-name CDS. If the bond defaults, investors expect to receive a proportion of their holding back; this amount is given by the recovery rate. If the probability of default is p, then investors will receive the recovery rate amount after default. Prior to this, they will receive $(1-p)$. To gain an accurate measure of the basis therefore, we should incorporate this probability of default factor.

7.3.4 Adjusted Z-spread

For the most accurate measure of the basis when undertaking relative value strategy, investors should employ the adjusted Z-spread, sometimes known as the C-spread. With this methodology, cash flows are adjusted by their probability of being paid. Since the default probability will alter over time, a static value cannot be used so the calculation is done using a binomial approach, with the relevant default probability for each cash flow pay-date. An adjusted Z-spread can be calculated by either converting the bond price to a CDS-equivalent price, or converting a CDS spread to a bond-equivalent price. This is then subtracted from the CDS spread to give the adjusted basis.

7.3.5 Adjusted Basis Calculation

We look now at pricing the CDS on the bond-equivalent convention; that is, converting the CDS price to a CDS-equivalent bond spread. This approach reduces uncertainty as there is a directly observable credit curve, the CDS curve, to which we can attach specific default probabilities. The CDS curve for a large number of reference names is liquid across the term structure from 1 through to 10, 15 and often 20 years. This is not the case with a corporate bond curve.[4]

The basis measure calculated using this method is known as the adjusted basis. It is given by

$$\text{Adjusted basis} = \text{Adjusted CDS spread} - \text{Z-spread} \qquad (7.3)$$

---

[4]Unless the corporate issuer in question has a continuous debt programme that means it has issued bonds at regular (say 6-month) intervals across the entire term structure.

The Z-spread is the same measure described in section 7.3 above, that is, the actual Z-spread of the cash bond in question. Instead of comparing it to a market CDS premium though, we compare it to the CDS-equivalent spread of the traded bond. The adjusted CDS spread is known as the adjusted Z-spread, CDS spread or C-spread. The adjusted basis is the difference between these two spreads.

The adjusted CDS spread is calculated by applying CDS market default probabilities to the cash bond in question. In other words, it gives an hypothetical price for the actual bond. This price is not related to the actual market price but should, unless the latter is significantly mis-priced, be close to it. It is a function of the default probabilities implied by the CDS curve, the assumed recovery rate and the bonds cashflows. Once calculated, the adjusted CDS spread is used to obtain the basis.

To calculate the adjusted CDS spread we carry out the following:

- Plot a term structure of credit rates using CDS quotes from the market. The quotes will all be single-name CDS prices for the reference credit in question. For the greatest accuracy we use as many CDS prices as possible (from the 6-month to the 20-year maturity, at 6-month intervals where possible), although the number of quotes will depend on the liquidity of the name.
- From the credit curve we derive a cumulative default probability curve. This is used together with the value for the recovery rate to construct a survival curve by bootstrapping default probabilities.
- The survival curve is used to price the bond in line with a binomial tree model.
- The adjusted CDS spread is the spread above (or below) the swap curve that equates the bond's observed market price to its hypothetical price calculated above.

The adjusted CDS spread can be viewed as the Z-spread of the bond at its hypothetical price. Comparing it to the actual Z-spread then gives us a like-for-like comparison and a more robust measure of the basis.

In other words, we produce an adjusted Z-spread based on CDS prices and compare it to the Z-spread of the bond at its actual market price. We have still compared the synthetic asset price to the cash asset price, as given originally in equation (7.1), but the comparison is now a logical one.

7.3.6 Illustration

The foregoing highlights some of the issues involved in measuring the basis. Irrespective of the calculation method employed, the one

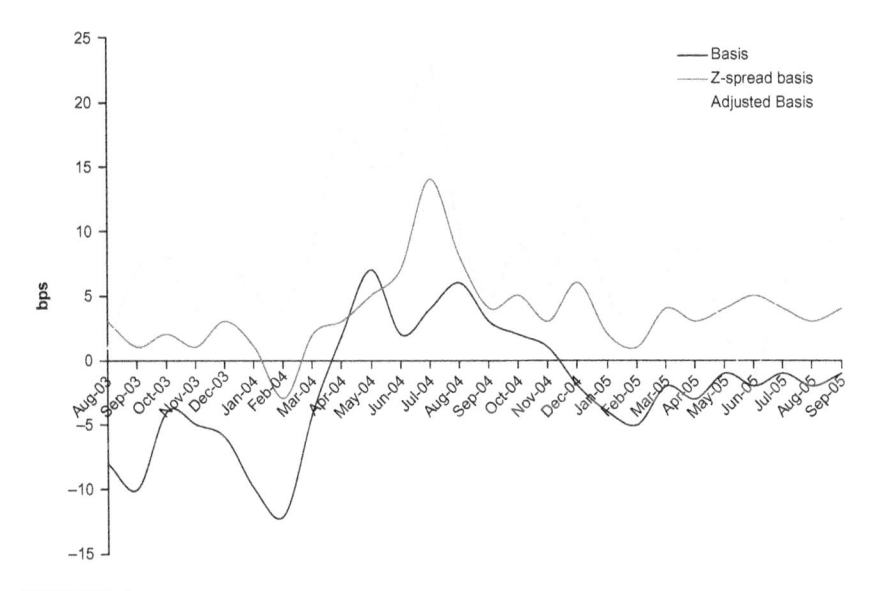

FIGURE 7.4 Selected telecoms corporate name, cash-CDS basis 2003–2005. *Yield source: Bloomberg. Calculation: Author.*

certainty is that corporate names in cash and synthetic markets do trade out of line; the basis is driven across positive and negative ranges due to the impact of a diverse range of technical, structural and market factors. Of course, any basis value that is not zero implies potential value in arbitrage trades in that reference name, be it buying or selling the basis.

The conventional situation is a positive basis, as a negative basis implies a risk-free arbitrage profit is available.

Market observation suggests that a negative basis is more common than might be expected however. Figure 7.4 shows the basis – measured in each of the three ways we have highlighted – for a selected corporate name in the telecoms sector for the period from August 2003 to September 2005. There is occasional relatively wide divergence between the measures, but all measures show the move from positive to negative territory and then back again.

7.4 MARKET OBSERVATIONS

The value of the CDS basis as a relative indicator became apparent, if it was not already, in the aftermath of the 2008 financial market crash. As basis spreads turned to negative, it was clear that the higher

134　　7. THE ASSET SWAP–CREDIT DEFAULT SWAP BASIS

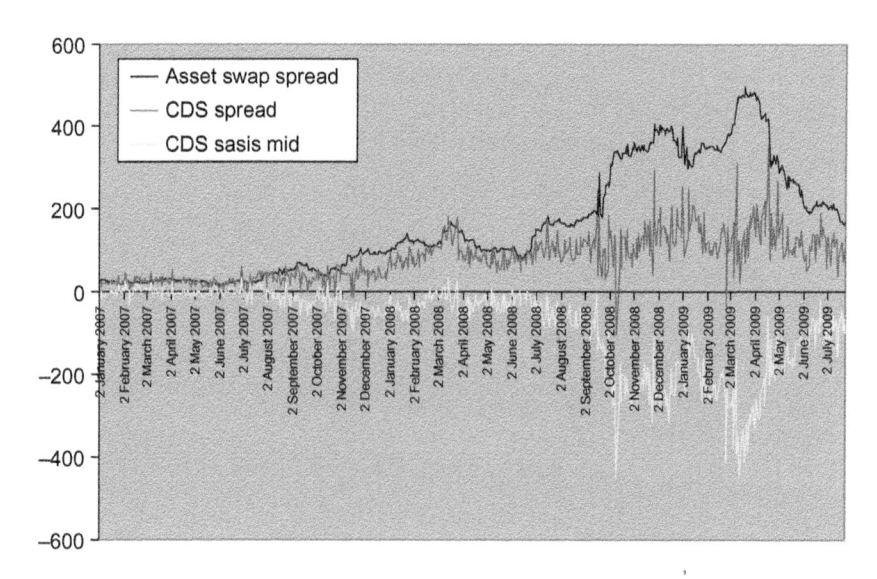

FIGURE 7.5 JPMorgan Chase CDS, ASW and basis levels 2007–2009. *Source: Bloomberg L.P.*

value of actual cash that prevailed, and still prevails at the time of writing, since the crash means that all else being equal the basis should be negative. In other words, investing in the cash bond should pay a higher return than simply writing CDS protection on the same name.

We illustrate this phenomenon first with Figure 7.5, which plots the CDS premium, asset swap spread and cash-CDS basis for JPMorgan Chase for the period 2007–2009, either side of the financial market crash of 2008. It is immediately apparent how the basis relationship changed after the crisis period. From a basis that hovered close to zero and moved between positive and negative, it turned quite obviously negative following the crash.

The change in basis relationship is, as one would expect, not confined to banks either although with some corporates the picture is more complicated. Figure 7.6(a) shows the position of individual bond yields and the continuous CDS term structure for Caterpillar Inc., a US industrial machinery manufacturer, in July 2008. We see that generally the basis is positive. Looking at the company's yields as at December 2008 (see Figure 7.6(b)) the position has reversed and the basis is now firmly negative. However, by January 2010 the picture is less clear, with bond yields lying both above and below the CDS curve, as shown at Figure 7.6(c).

Clearly the cash–synthetic relationship is complex and one that is driven by a number of different factors. Generally, if we hold

7.4 MARKET OBSERVATIONS 135

(a)

CDS and asset swap spreads - Caterpillar

(b)

CDS and asset swap spreads - Caterpillar

(c)

CDS and asset swap spreads - Caterpillar

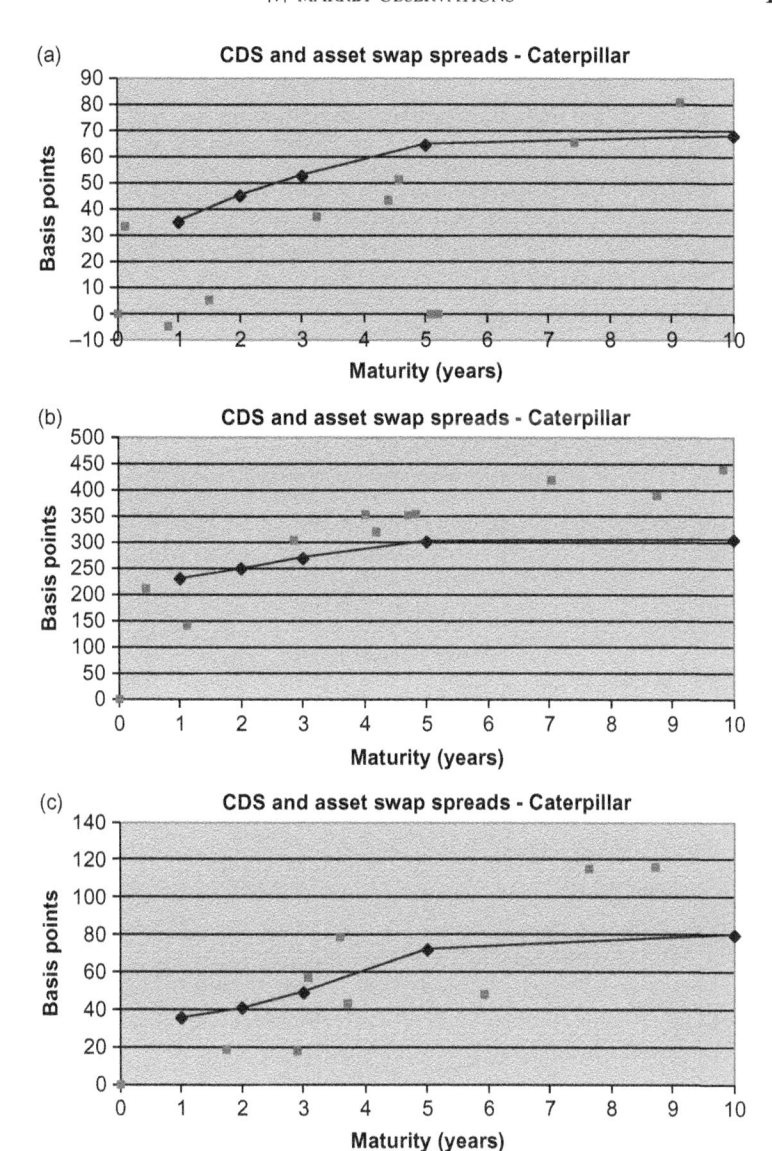

FIGURE 7.6 (a) Caterpillar bond yields and CDS spread term structure, July 2008. (b) Caterpillar bond yields and CDS spread term structure, December 2008. (c) Caterpillar bond yields and CDS spread term structure, January 2010. *Source of (a-c): Bloomberg L.P.*

all else equal and constant, the major influence is the premium attached to actual cash. However, this factor interacts with a number of other determinants to drive the ultimate basis value at any one time.

7.5 THE ITRAXX INDEX BASIS

The iTraxx and CDX synthetic credit indices are a reasonably liquid set of CDS contracts and as such are viewed as a key credit market indicator by investors. As a market benchmark, they can be taken to represent the market as a whole, similar to an equity index, and as such can provide a guide to the health of the credit market as well as imply market perception of future direction. Because there is an intrinsic 'fair value' implied for the index, which differs from the actual market level at any time, there is also a CDS index basis that can be observed. Assessing this basis therefore is an important part of investor's relative value analysis of the market.

A CDS index basis arises because of the different approaches available to valuing the index spread.[5] We can describe the following:

- Theoretical spread: this is the average of the single-name CDS spreads for all the constituent names in the index, weighted by probability of default.[5] It is also known as the *fair spread*.
- Market spread: this is the spread quoted for trading in the market. It is based on the theoretical spread, but adjusted using a flat credit curve. It is also known as the *intrinsic spread* or the *real spread*.

There is also the *average spread*, which is simply the average of all the constituent name CDS spreads equally weighted.

The index basis is given by:

$$\text{Basis} = \text{Market spread} - \text{Theoretical spread}$$

From a relative value analysis point of view we require a tractable means of calculating the two spreads or at least of measuring the basis. This is considered next.

In practice the basis in index CDS is small, with exceptions being observed for high-volatility indices. We consider the June 2011 iTraxx investment-grade Europe index (series 5), which commenced trading in March 2006 as the 5-year benchmark. Figure 7.7 shows the levels for the market and theoretical spreads for this index from March 2006 to June 2006, while Figure 7.8 shows the basis. Observe that the basis fluctuates from negative to positive and back, although the absolute values are small.

Irrespective of its absolute value, observing the basis enables investors to gauge perceived fair value for the index. By comparing the market spread against the theoretical spread we can assess whether the index is trading at fair value or not. For instance, during June 2005 the iTraxx Europe Cross-over 5-year index was observed with a basis of

---

[5]Probability of default (PD) = Spread/$(1 - R)$ where R is the recovery rate.

7.5 THE ITRAXX INDEX BASIS

FIGURE 7.7 iTraxx IG Europe index CDS theoretical and market spreads, Mar–Jun 2006. *Source: KBC Financial Products.*

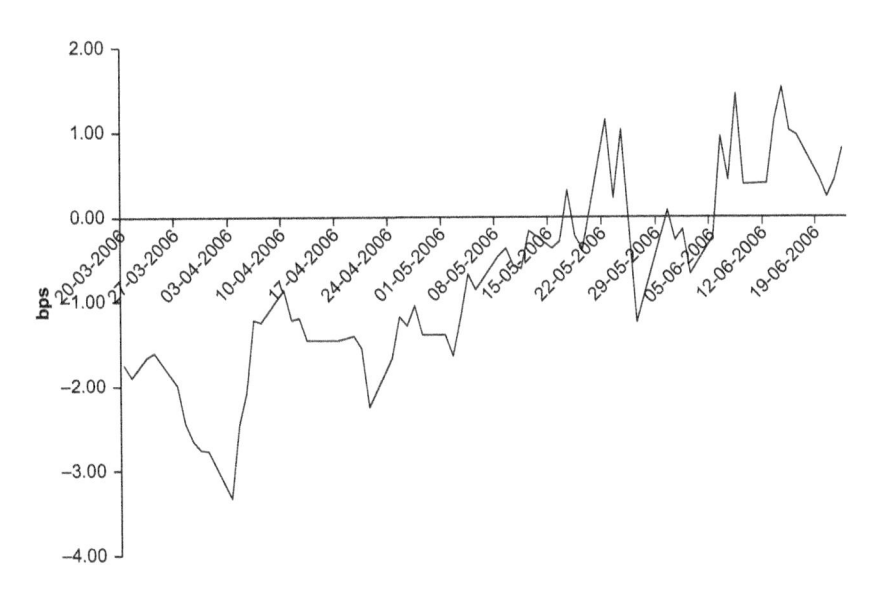

FIGURE 7.8 iTraxx IG Europe index CDS basis. *Source: KBC Financial Products.*

138   7. THE ASSET SWAP–CREDIT DEFAULT SWAP BASIS

TABLE 7.1 Index CDS Basis for Selected Indices, June 2006

Index	Average Spread	Market Spread	Theoretical Spread	Average Spread Basis	Theoretical Spread Basis
iTraxx IG Europe 5-year	44	43	44	−1	−1
iTraxx IG Europe High Vol 5-year	81	83	81	+2	+2
iTraxx Cross-over Europe 5-year	349	350	337	+1	+13
CDX NA IG 5-year	63	59	61	−4	−2
CDX NA IG High Vol 5-year	139	133	133	−6	0

Source: MarkIt

around 13 basis points, which suggested mis-pricing of the index fair value. For comparison we show the basis values for other indices at the same time in Table 7.1, with data reported by MarkIt partners.

7.6 NEGATIVE BASIS TRADE: CHECKING THE THEORETICAL BOND PRICE

We have noted in this chapter the various means by which the basis can be measured, and the approach to conducting the requisite analysis before putting on the basis trade. CDS market-making banks often make a basis package available from the outset; when this happens there is no need for the arbitrager to construct the trade because both the cash and the CDS sides will take place simultaneously.

For example, on 30 September 2009 a market maker offered the following maturity matched negative basis package:

- EUR15 million of British Telecom 6.50% July 2015 bonds, rated BBB and Baa2, trading at a negative basis of 56 bps.

The indicators shown were:

- Price   107.83;
- Yield   4.90%;
- Spread   246 bps;
- Z-spread   206;
- CDS   150;
- Basis   56.

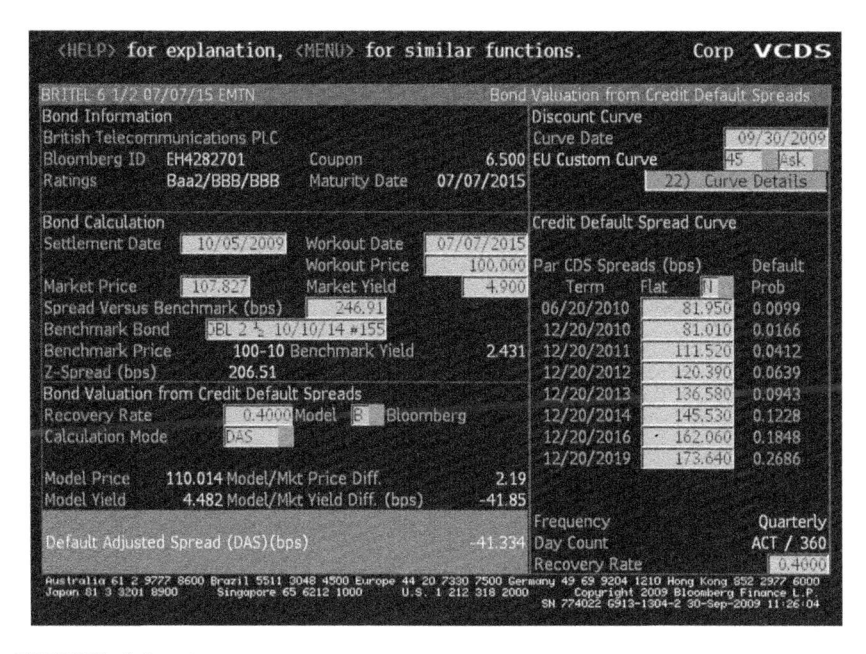

FIGURE 7.9 Bloomberg screen used to analyse negative basis package in British Telecom 6.5% 2015 bond, as at 30 September 2009. Note CDS-implied price of 110.01. © Bloomberg L.P. Used with permission. Visit www.bloomberg.com

The decision to trade will reflect the individual's view; however, it is possible to check the theoretical bond price derived from the CDS curve using Bloomberg screen VCDS. This is shown in Figure 7.9. We see that the using a recovery rate of 40% the theoretical bond price of 110.01 is some way away from the market price of 107.83. Consequently we conclude that the market perception of cash market value is different from that implied by the synthetic market; hence the relatively large negative basis. On this analysis, as the cash bond is 'cheap' relative to the theoretical price, the negative basis trade has some attraction, as the trader will be buying the cash and hedging the credit risk via the CDS market.

References

Choudhry, M. (2001). *Some issues in the asset-swap pricing of CDSs*. Derivatives Week, Euromoney Publications, 5 December.

Choudhry, M. (2004). *Fixed income markets: instruments, applications, mathematics*. Singapore: John Wiley and Sons.

Choudhry, M (2006). *The credit default swap basis*. Princeton, NJ: Bloomberg Press.

Chertsey, 2010, including Phil Broadhurst

Phil Broadhurst, Contributor to First Edition, Reading, 1988

Glossary

Don't get lost in all that Stock Exchange stuff...keep the faith, the bass guitar and the Fred Perry.
—Phil Broadhurst, *The Mighty Utterance*, 1989

Amortizing A financial instrument whose nominal principal amount decreases in size during its life.

Arbitrage The process of buying securities in one country, currency or market, and selling identical securities in another to take advantage of price differences. When this is carried out simultaneously, it is in theory a risk-free transaction. There are many forms of arbitrage transactions. For instance, in the cash market a bank might issue a money market instrument in one money centre and invest the same amount in another centre at a higher rate, such as an issue of 3-month US dollar CDS in the United States at 5.5% and a purchase of 3-month Eurodollar CDS at 5.6%. In the futures market, arbitrage might involve buying 3-month contracts and selling forward 6-month contracts.

Arbitrage CDO A collateralized debt obligation (CDO) that has been issued by an asset manager and in which the collateral is purchased solely for the purpose of securitizing it to exploit the difference in yields ('arbitrage') between the underlying market and the securitization market.

Asset & liability management (ALM) The practice of matching the term structure and cash flows of an organisation's asset and liability portfolios to maximise returns and minimise risk.

Asset-backed securities Securities that have been issued by a special purpose legal entity (SPV) and are backed by principal and interest payments on existing assets, which have been sold to the SPV by the deal originator. These assets can include commercial bank loans, credit card loans, auto loans, equipment lease receivables, and so on.

Asset swap An interest-rate swap or currency swap used in conjunction with an underlying asset such as a bond investment.

Average life The weighted-average life of a bond, the estimated time to return principal based on an assumed prepayment speed. It is the average number of years that each unit of unpaid principal remains outstanding.

Balance sheet CDO A CDO backed by a static pool of assets that were previously on the balance sheet of the originator.

Basis The underlying cash market price minus the futures price. In the case of a bond futures contract, the futures price must be multiplied by the conversion factor for the cash bond in question.

Basis points In interest-rate quotations, 0.01%.

Basis swap An interest-rate swap where both legs are based on floating rate payments.

Bullet A loan/deposit has a bullet maturity if the principal is all repaid at maturity (see **amortizing**).

CDO Collateralized debt obligation, a structured financial product.

CMBS Commercial mortgage-backed securities.

Credit default swaps Agreement between two counterparties to exchange disparate cash flows, at least one of which must be tied to the performance of a credit-sensitive asset or to a portfolio or index of such assets. The other cash flow is usually tied to a floating-rate index (such as Libor) or a fixed rate or is linked to another credit-sensitive asset.

Credit derivatives Financial contracts that involve a potential exchange of payments in which at least one of the cash flows is linked to the performance of a specified underlying credit-sensitive asset or liability.

Credit enhancement A level of investor protection built into a structured finance deal to absorb losses among the underlying assets. This may take the form of cash, 'equity' subordinated note tranches, subordinated tranches, cash reserves, excess spread reserve, insurance protection ('wrap'), and so on.

Credit risk (or default risk) The risk that a loss will be incurred if a counterparty to a derivatives transaction does not fulfil its financial obligations in a timely manner.

Credit risk (or default risk) exposure The value of the contract exposed to default. If all transactions are marked to market each day, such positive market value is the amount of previously recorded profit that might have to be reversed and recorded as a loss in the event of counterparty default.

Credit spread The interest-rate spread between two debt issues of similar duration and maturity, reflecting the relative creditworthiness of the issuers.

Currency swap An agreement to exchange a series of cash flows determined in one currency, possibly with reference to a particular fixed or floating interest payment schedule, for a series of cash flows based in a different currency (see **interest-rate swap**).

Default correlation The degree of covariance between the probabilities of default of a given set of counterparties. For example, in a set of counterparties with positive default correlation, a default by one counterparty suggests an increased probability of a default by another counterparty.

Diversity score A Moody's CDO calculation that assigns a numeric value to an asset portfolio that represents the number of uncorrelated assets theoretically in the portfolio. A low diversity score indicates industry and/or geographical concentration, and will be penalised in the ratings process.

Equity-linked swap Swap where one of the cash flows is based on an equity instrument or index, when it is known as an equity index swap.

Equivalent life The weighted average life of the principal of a bond where there are partial **redemptions**, using the **present values** of the partial redemptions as the weights.

Excess spread Total cash left over in a securitisztion, after paying all costs.

Expected (credit) loss Estimate of the amount a derivatives counterparty is likely to lose as a result of default from a derivatives contract, with a given level of probability. The expected loss of any derivative position can be derived by combining the distributions of credit exposures, rate of recovery and probabilities of default.

Expected default rate Estimate of the most likely rate of default of a counterparty expressed as a level of probability.

Expected rate of recovery See **rate of recovery**.

Face value The principal amount of a security generally repaid ('redeemed') all at maturity, but sometimes repaid in stages, on which the **coupon** amounts are calculated.

Floating rate An interest-rate set with reference to an external index. Also, an instrument paying a floating rate is one where the rate of interest is refixed in line with market conditions at regular intervals, such as every 3 or 6 months. In the current market, an exchange rate determined by market forces with no government intervention.

Floating rate note Capital market instrument on which the rate of interest payable is refixed in line with market conditions at regular intervals (usually 6 months).

Forward In general, a deal for value later than the normal value date for that particular commodity or instrument. In the foreign exchange market, a forward price is the price quoted for the purchase or sale of one currency against another where the value date is at least one month after the spot date.

GLOSSARY

Future value The amount of money achieved in the future, including interest, by investing a given amount of money now (see **time value for money, present value**).

Futures contract A deal to buy or sell some financial instrument or commodity for value on a future date. Unlike a **forward** deal, futures contracts are traded only on an exchange (rather than OTC), have standardized contract sizes and value dates, and are often only contracts for differences rather than deliverable.

GIC Guaranteed investment contract. A bank account that pays either a fixed rate for its life, or a fixed spread under Libor for its life.

Hedging Protecting against the risks arising from potential market movements in exchange rates, interest rates or other variables.

Historic volatility The actual **volatility** recorded in market prices over a particular period.

Implied volatility The volatility used by a dealer to calculate an option price; conversely, the volatility implied by the price actually quoted.

Interbank The market in unsecured lending and trading between banks of roughly similar credit quality.

Interest rate swap An agreement to exchange a series of cash flows determined in one currency, based on fixed or **floating** interest payments on an agreed **notional** principal, for a series of cash flows based in the same currency but on a different interest rate. May be combined with a **currency swap.**

Internal rate of return The yield necessary to discount a series of cash flows to a net present value of zero.

Liability swap An interest-rate swap or currency swap used in conjunction with an underlying liability such as a borrowing (see **asset swap**).

Libid The London Interbank Bid Rate, the rate at which banks will pay for funds in the interbank market.

Libor The London Interbank Offered Rate, the lending rate for all major currencies up to 1-year set at 11 am each day by the British Bankers Association.

Liquidity A word describing the ease with which one can undertake transactions in a particular market or instrument. A market where there are always ready buyers and sellers willing to transact at competitive prices is regarded as liquid. In banking, the term is also used to describe the requirement that a portion of a bank's assets be held in short-term risk-free instruments, such as government bonds, T-Bills and high quality Certificates of Deposit.

Long A long position is a surplus of purchases over sales of a given currency or asset, or a situation which naturally gives rise to an organization benefitting from a strengthening of that currency or asset. To a money market dealer, however, a long position is a surplus of borrowings taken in over money lent out (which gives rise to a benefit if that currency weakens rather than strengthens) (see **short**).

Long-term assets Assets which are expected to provide benefits and services over a period longer than 1 year.

Long-term liabilities Obligations to be repaid by the firm more than 1 year later.

Market-maker Market participant who is committed, explicitly or otherwise, to quoting two-way bid and offer prices at all times in a particular market.

Market risk Risks related to changes in prices of tradable macroeconomics variables, such as exchange rate risks.

Mark-to-market The act of revaluing securities to current market values. Such revaluations should include both coupon accrued on the securities outstanding and interest accrued on the cash.

Maturity date Date on which stock is redeemed.

Net present value The net present value of a series of cash flows is the sum of the present values of each cash flow (some or all of which maybe negative).

Nominal amount Same as **face value** of a security.

Non-performing A loan or other asset that is no longer being serviced, or has experienced default.

Notional In a bond futures contract the bond bought or sold is a standardized non-existent notional bond, as opposed to the actual bonds which are deliverable at maturity. Contracts for differences also require a notional principal amount on which settlement can be calculated.

NPV See **net present value.**

OTC Over-the-counter. Strictly speaking, any transaction not conducted on a registered stock exchange. Trades conducted via the telephone between banks, and contracts such as forward rate agreements (FRAs) and (non-exchange traded) options, are said to be 'over-the-counter' instruments. OTC also refers to non-standard instruments or contracts traded between two parties; for example, a client with a requirement for a specific risk to be hedged with a tailor-made instrument may enter into an OTC structured option trade with a bank that makes markets in such products.

Over-the-counter An OTC transaction is one dealt privately between any two parties, with all details agreed between them, as opposed to one dealt on an exchange – for example, a **forward** deal as opposed to a **futures contract.**

Paper Another term for a bond or debt issue.

Par When the price of a security is equal to the face value, usually expressed as 100, it is said to be trading at par. A par swap rate is the current market rate for a fixed **interest-rate swap** against **Libor.**

Par yield curve A curve plotting maturity against **yield** for bonds priced at par.

Plain vanilla See **vanilla.**

Present value The amount of money that needs to be invested now to achieve a given amount in the future when interest is added (see **time value for money, future value**).

Primary market The market for new debt, into which new bonds are issued. The primary market is made up of borrowers, investors and the investment banks which place new debt into the market, usually with their clients. Bonds that trade after they have been issued are said to be part of the secondary market.

Quanto swap A swap where the payments on one or both legs are based on a measurement (such as the interest rate) in one currency but payable in another currency.

Rate of recovery Estimate of the percentage of the amount exposed to default – i.e. the credit risk exposure – that is likely to be recovered by an institution if a counterparty defaults.

Record date A coupon or other payment due on a security is paid by the issuer to whoever is registered on the record date as being the owner.

Redeem A security is said to be redeemed when the principal is repaid.

Redemption yield The rate of interest at which all future payments (coupons and redemption) on a bond are discounted so that their total equals the current price of the bond (inversely related to price).

Return on equity The net earning of a company divided by its equity.

Secondary market The market in instruments after they have been issued. Bonds are bought and sold after their initial issue by the borrower, and the marketplace for this buying and selling is referred to as the secondary market. The new issues market is the **primary market.**

Securitization Suresh Sundaresan has defined securitization as a framework in which some illiquid assets of a corporation or a financial institution are transformed into a package of securities backed by these assets, through careful packaging, credit enhancements, liquidity enhancements and structuring.

Security A financial asset sold initially for cash by a borrowing organization (the 'issuer'). The security is often negotiable and usually has a maturity date when it is redeemed.

GLOSSARY

Short A short position is a surplus of sales over purchases of a given currency or asset, or a situation which naturally gives rise to an organization benefitting from a weakening of that currency or asset. To a money market dealer, however, a short position is a surplus of money lent out over borrowings taken in (which gives rise to a benefit if that currency strengthens rather than weakens) (see **long**).

Special A security which for any reason is sought after in the repo market, thereby enabling any holder of the security to earn incremental income (in excess of the general collateral rate) through lending them via a repo transaction. The repo rate for a special will be below the GC rate, as this is the rate the borrower of the cash is paying in return for supplying the special bond as collateral. An individual security can be in high demand for a variety of reasons, for instance if there is sudden heavy investor demand for it or (if it is a benchmark issue) it is required as a hedge against a new issue of similar maturity paper.

Synthetic A package of transactions which is economically equivalent to a different transaction. In the structured finance market, a transaction that replicates some of the economic effects of a cash securitization without recourse to an actual sale of assets, and which involves the use of credit derivatives.

Time value for money The concept that a future cash flow can be valued as the amount of money which it is necessary to invest now in order to achieve that cash flow in the future (see **present value, future value**).

Underlying The underlying of a futures or option contract is the commodity or financial instrument on which the contract depends. Thus the underlying for a bond option is the bond; the underlying for a short-term interest-rate futures contract is typically a 3-month deposit.

Underwriting An arrangement by which a company is guaranteed that an issue of debt (bonds) will raise a given amount of cash. Underwriting is carried out by investment banks, who undertake to purchase any part of the debt issue not taken up by the public. A commission is charged for this service.

Unexpected default rate The distribution of future default rates is often characterized in terms of an expected default rate (e.g. 0.05%) and a worst-case default rate (e.g. 1.05%). The difference between the worst-case default rate and the expected default rate is often termed the **'unexpected default'** (i.e. 1% = 1.05 − 0.05%).

Unexpected loss The distribution of credit losses associated with a derivative instrument is often characterized in terms of an **expected loss** or a **worst-case loss**. The unexpected loss associated with an instrument is the difference between these two measures.

Value-at-risk (VAR) Formally, the probabilistic bound of market losses over a given period of time (known as the holding period) expressed in terms of a specified degree of certainty (known as the confidence interval). Put more simply, the VAR is the **worst-case** loss that would be expected over the holding period within the probability set out by the confidence interval. Larger losses are possible, but with a low probability. For instance, a portfolio whose VAR is $20 million over a 1-day holding period, with a 95% confidence interval, would have only a 5% chance of suffering an overnight loss greater than $20 million.

Value date The date on which a deal is to be consummated. In some bond markets, the value date for coupon accruals can sometimes differ from the settlement date.

Vanilla A vanilla transaction is a straightforward one.

VAR See **value-at-risk**

Variance−covariance methodology Methodology for calculating the **value-at-risk** of a portfolio as a function of the **volatility** of each asset or liability position in the portfolio and the correlation between the positions.

Volatility The standard deviation of the continuously compounded return on the underlying. Volatility is generally annualized (see **historic volatility, implied volatility**).

Worst-case loss Estimate of the largest amount a derivative counterparty is likely to lose, with a given level of probability, as a result of default from a derivatives contract or portfolio.

Yield The interest rate that can be earned on an investment, currently quoted by the market or implied by the current market price for the investment – as opposed to the coupon paid by an issuer on a security, which is based on the coupon rate and the face value. For a bond, generally the same as yield to maturity unless otherwise specified.

Yield curve Graphical representation of the maturity structure of interest rates, plotting yields of bonds that are all of the same class or credit quality against the maturity of the bonds.

SPG in Liverpool, 1986

    I first met Johnny when I threw him across a table at a Uni disco while dancing to SLF's 'Suspect Device'. If I'd known then that he was a schoolboy boxing champion I don't think I'd have bothered!

    That same night we had our first joint 'confrontation', with a college grebo who didn't like the idea of us pogoing over him — a quick fist taught him a thing or two!

    We've battered the NF together, we were the only anti-nazis at a Business gig full of NF (did we _____ it or what?!), we took on 9 Yanks in Earls Court tube station but did we run?

    Many gigs, laughs and my pre-occupation with work later, we find ourselves in different lands and with different tastes in music, but the spirit of 1985 still lives on — it was a shame that Stuart Morrow left New Model Army and that the Villa never won anything.

    Keep on Keepin' On till the fight is won!

Nik Slater
SPG
June 1993

    What do you think of your bass player? I hear he's had quite a chequered career musically?

    Firstly, my bass player is a complete _____ for leaving SPG the first time back in 1986. Secondly, he is a flippin' great bass player, see him play live and you will know what I mean. He supports some dodgy south London football team though so he isn't that mentally stable. As far as music goes, he appeared with *The New English* on BBC TV and now sessions for some group who are destined for big things, or so he claims.

    He is probably one of the best bass players in the UK today, he has so much energy live that if you could hook it to the National Grid, you could run London for a week.

Nik Slater
This is a load of _____ so don't buy it fanzine
February 1990

Index

A
Air Products & Chemicals, 128f
Alpha-Sires, 58–61
Arbitrage
  asset swap/CDS price differentials, 21, 125–127
  definition, 143–144
Asia, crises, 6
Asset swaps, 22–23, 28–30, 69, 88, 96–98, 123
  arbitrage opportunities, 125–127
  Bloomberg screens, 30, 128f, 129f
  CDSs, 123
  concepts, 22–23, 28–30, 69, 88, 96–98, 123
  credit events, 30–32, 39–40
  examples, 30, 34b, 96–98, 125
  generic details, 29, 29f, 123–125
  prices, 88, 96–98, 123–127
  TRSs, 30–36
Asset transfers, banks, 5, 18–19, 33–34

B
Balance sheets
  credit derivatives, 19, 33–34
  credit ratings, 7
  off-balance sheet instruments, 19, 33–34, 75–76, 78–79
  TRSs, 78–79
Banco Bilbao Vizcaya Argentaría (BBVA), 25f
Bankruptcies, credit events, 20, 49
Banks
  asset transfers, 5, 18–19, 33–34
  Barings, 11
  cash settlements, 41
  CLNs, 45, 53–64
  credit derivatives, 1, 6, 17–21, 33–34, 36, 40–41, 44, 53, 61
  credit ratings, 10
  Libor, 3, 11, 27–30, 32–34, 76–77, 96–97, 145
Barings, 11
Basket credit-linked notes, concepts, 62
Basket default swaps, concepts, 26–27

BBVA *see* Banco Bilbao Vizcaya Argentaria
Beneficiaries, TRSs, 33–34
Black–Scholes–Merton model, 88, 93, 94t, 95t
Bloomberg screens, 9f, 31f, 36, 38–39, 47, 47f, 48f, 56–58, 57f, 58f, 59f, 60f, 61f, 71, 72f, 109f, 139f, 140
Bond derivatives, prices, 33–34
Bonds, 6, 45, 78
  asset swaps, 96–98, 123–127
  cash flows, 28–32, 35, 77
  cash/synthetic position comparisons, 45
  CLNs, 45, 53–64
  cost of funds, 22–23
  credit spread risks, 7–9, 19, 23–24, 89
  euro, 12, 24
  risk-free bonds, 56
  risks, 5–6, 10, 29, 42, 44–45, 52, 145
  special bonds, 147
British Petroleum plc, 11
British Telecom plc, 30, 31f, 57–58, 58f, 59f, 71, 71t, 138
Broadhurst, P., 138t

C
Canada, 46
Capital structure arbitrage, TRSs, 76–77
Carry, TRSs, 75–76
Cash flows
  asset swaps, 28–30
  bonds, 28–30
  CDSs, 101–104
  credit ratings, 7
  TRSs, 30–32, 76
Cash settlements, 22, 24, 40–42, 41f, 45, 53–54, 63, 108–109
  concepts, 41–42, 53–54, 63, 108–109
  mechanisms, 43–44
CDOs *see* Collateralised debt obligations
CDSs *see* Credit default swaps
Cheapest-to-deliver concepts, 20–21, 41–42
CLNs *see* Credit-linked notes
'Collateral', CLNs, 54–55, 61–62, 62f, 77

149

INDEX

Collateralised debt obligations (CDOs), 54, 61–64
  CLNs, 61–64
  concepts, 54, 61–62, 143
Commercial banks, 18–19, 65, 67, 76, 143
Commercial paper, 10–11
Conseco case, 44–45
Contingent cash-flow leg, CDSs, 104
Corporate Bonds and Structured Financial Products (author), 54–55, 65
Cost of funds, 22–23
Credit analysis, concepts, 10–12, 67
Credit card issuers, CLNs, 56–57
Credit curves, CDSs, 46, 98–100, 99f, 131–132, 136
Credit default risks, concepts, 7
Credit default swaps (CDSs), 2–3, 5–6, 22–28
  arbitrage opportunities, 125–127
  asset swaps, 123–127
  basket CDS, 69–71
  basket default swaps, 26–27
  Bloomberg screens, 30
  buying or selling protection, 27
  calculator, 108–109
  cash flows, 101–104
  CLNs, 53–64
  concepts, 2–3, 5–6, 22–28
  credit curves, 46, 98–100, 99f, 131–132, 136
  EDS contrasts, 29f
  examples, 26b, 44–48, 125, 132–133
  generic price page, 124
  market pricing-approach, 104–106
  maturity issues, 24, 98–100
  prices, 89, 98, 101
  risks, 44–45, 83–84, 98–100, 125–127
  structure, 23–26
  term-sheet sample, 50b
  theoretical pricing approach, 101–104
  unintended risks, 44
  'unwinding' a position, 27–28
  see also Unfunded credit derivatives
Credit derivatives
  applications, 17–19, 65
  banks, 17–19, 22–23, 30–34, 75–76
  Bloomberg screens, 9f, 31f, 36, 38–39, 47, 47f, 48f, 56–58, 57f, 58f, 59f, 60f, 61f, 71, 72f, 109f, 139f, 140
  characteristics, 21–23
  concepts, 17
  credit pair trade, 68–69
  credit risks, 17–21, 65, 123–127
  flexibility benefits, 17–19
  generic details, 21–22
  insurance policies, 22
  ISDA definitions, 20–21, 34, 44–47, 49b, 52t, 78, 108–109
  legal documentation, 20–21
  maturity issues, 24, 30–32, 40–41, 75–76, 98
  payout conditions, 17–28, 40–44, 51t, 54, 63, 101–104
  pricing, 106–108
  reasons, 18, 65
  relative value trading of, 67–71
  risks, 17–21, 44–45, 65–67, 123–127
  selection of, 68
  settlement, 17–28, 40–44, 51t, 54, 63, 101–104
  transparency benefits, 17
  see also Funded...; Prices; Unfunded...
Credit events, 2–3, 13–14, 17–18
  asset swaps, 28
  CLNs, 54–61
  concepts, 2–3, 13–14, 17–18
  definitions, 20–21, 44
  ISDA definitions, 20–21, 44, 62–63
  risks, 44
  verification agents, 40
Credit evolutionary path, price factors, 87
Credit exposure, portfolios, 83–84
Credit options, concepts, 56–57
Credit ratings, 1, 20, 28–30, 33, 44–45, 54–56, 90, 101–102, 111
  agencies, 7, 8t, 10–12, 44, 76, 101–102
  analysis considerations, 7, 10
  CLNs, 56–61, 63–64
  concepts, 1, 20, 126
  cost bearers, 11
  credit spread risks, 7–9, 20, 23–24
  credit watch, 12
  FtD CLNs, 26, 62–64, 63f, 69
  interest rates, 11
  Libor, 28–29
  purposes, 11
  scales, 1, 20, 23–24, 126
  transition matrices, 89–90
Credit risks, 1, 63–65, 123–127, 144
  analysis considerations, 10
  banks and, 4–6
  concepts, 1, 65, 144
  credit derivatives, 17–21, 65, 123–127
  management issues, 65

market risks, 10
name recognition, 11
portfolios, 10, 18–19, 44–45, 64–65
types, 7–9
value-at-risk measurement, 147
yield links, 10
Credit spread options
    concepts, 90, 102
Credit spread products, 94–98
Credit spread risks
    concepts, 7–9
    credit ratings, 10
    government bonds, 7–9, 85
    models, 92–94
    yield, 7–9, 94–96
Credit swaps *see* Credit default swaps
Credit switches, portfolios, 84
Credit watch, concepts, 12
Credit-linked notes (CLNs), 22, 45, 53–61, 83
    basket credit-linked notes, 62
    Bloomberg screens, 56*f*, 57*f*
    bonds, 54–56
    cash/synthetic position comparisons, 45
    CDOs, 61–64
    CDSs, 54–56, 61–62
    concepts, 22, 45, 53–61, 83
    credit ratings, 56–61, 63–64
    description, 54–56
    examples, 56–61
    forms, 56–57
    FtD CLNs, 26, 62–64, 63*f*, 69
    issuer types, 56–64
    maturity issues, 53, 61–62
    risks, 44–45, 64, 83–84
    structured products, 27, 40, 42, 54–55, 61–64, 83, 108
    see also Funded credit derivatives
Currency-linked notes, 85

D

Das, S., 28–29, 79–83, 89
Das–Tufano model (DT), prices, 90
Default swap curves *see* Credit curves
Default swaps *see* Credit default swaps
Deliverable obligations, concepts, 22, 40–42, 51*t*, 63–64
Digital credit derivatives, 41
Diversification uses, 5, 63–64, 79, 127–128
DT model *see* Das–Tufano model
Duffie–Singleton model, prices, 91–92, 104
Dynamic credit risks, 18

E

Equity default swaps (EDSs)
    generic structure, 29*f*
Equity-linked swap, 144
Equivalent life, 144
Euro, 12, 24, 71
Europe, restructuring events, 20–21
European options, 93

F

'Fair value' concepts, 71, 87, 90, 98, 136–138
First-to-default CLNs (FtD CLNs), 26, 62–64, 63*f*, 69
Fitch, 7, 8*t*, 11–12, 14*f*
Fixed amount contracts, 40–41
Fixed rates, asset swaps, 29, 123–124
Flexibility benefits, credit derivatives, 17–19, 54, 65, 67
Floating rate notes (FRNs), 124, 144
Floating rates
    asset swaps, 123–125
    concepts, 123–125, 144
Ford Motor, 60*f*, 61*f*
Forward credit spread, concepts, 94–96
Francis, J., 30
FRNs *see* Floating rate notes
FtD CLNs *see* First-to-default CLNs
Funded credit derivatives
    concepts, 19, 22–23, 33, 53, 61–63
    definition, 22
    TRSs, 33, 34*b*, 53, 61–63
    see also Credit-linked notes

G

General collateral rates, 77
Glossary, 143
'Go long/short', concepts, 23–24, 27, 85
Government bonds
    credit spread risks, 7–9
Guarantors, TRSs, 24, 32*f*, 33, 40–41

H

Hazard rates, concepts, 91–92
Hedging, 14–15, 18, 70–71, 123–124, 140, 145
High-yield bonds, pricing models, 88
Household Finance Corporation, 58, 59*f*, 60*f*
Hull–White model, prices, 102

I

Index swaps, concepts, 84–85
Insurance policies, credit derivatives, 6
Interest rates

Interest rates (*Continued*)
  credit ratings, 10–11, 17, 19
  risks, 22–23, 30–32, 68, 77
  see also Credit spread...
Interest-rate swaps, 22–23, 28–29, 34, 65, 67, 77, 110, 117, 145
International Swap Derivatives Association (ISDA), 20–21, 34, 44–47, 49b, 52t, 78, 108–109
Investment-grade credit ratings, 24
ISDA *see* International Swap Derivatives Association

J

Jarrow, Lando and Turnbull model (JLT), prices, 89–90
JLT *see* Jarrow, Lando and Turnbull model
JPMorgan, 47
JPMorgan approach, prices, 47, 134

L

Lando, D., 89
Legal documentation, credit derivatives, 20
Libid, 22–23, 75–76, 125, 145
Libor, 3, 11, 22–23, 27–30, 29f, 32–35, 32f, 58, 61–62, 66, 76–79, 83–85, 96–98, 123–124, 126
Liquidity issues
  CDSs, 19, 125–127
  concepts, 89–92, 125–127, 145
  reduced-form pricing models, 89–92
Longstaff, F.A., 102

M

Marked-to-market issues
  concepts, 33, 83, 145
  credit exposures, 83
  TRSs, 33
Market approach, prices, 104–106
Market makers, credit derivatives, 42–43, 47, 138, 145
Market requirements, settlement agreements, 42–43
Market risks, concepts, 68
Maturity issues
  CDSs, 24, 28, 75–76, 98
  CLNs, 53, 61–62
  credit derivatives, 24, 30–32, 75–76, 98
  TRSs, 30–33, 75–76
Mean-reversion model prices, credit spread modelling, 93–94
Medium-term notes, 10
Merrill Lynch, 58–61

Merton, R.C., 88
Microsoft, 11–12, 117
Modified restructuring, concepts, 20–21, 39–40, 45, 49–50
Moody's, 6–7, 8t, 11–14, 13t, 20, 44–45, 66, 101–102, 144

N

Name recognition, credit risks, 11
No-arbitrage pricing models, concepts, 89, 104, 110, 117, 122, 124

O

Obligations
  concepts, 7, 10, 13–14, 20–22, 27, 40–43, 45, 49
  ISDA definitions, 20, 49, 52
Off-balance sheet instruments, concepts, 19, 33–34, 78–79
Offer documents, 11
Operational risks, 6
Options
  credit spread options, 90, 102
  European options, 93
  pricing models, 89–92
OTC *see* Over-the-counter products
Over-the-counter products (OTC), 65, 146

P

Payout conditions, credit derivatives, 17–18, 41, 44, 63, 101
Physical settlements, 20–22, 24, 40–43, 41f, 51t, 53–54, 63
  concepts, 20–22, 24, 40–43, 41f, 51t, 53–54, 63
  mechanisms, 41–43, 51t, 54, 63
Portfolios, 4–5, 7
  applications, 75–79
  credit exposure, 75–77, 79–84, 123–124
  credit risks, 7–9, 45, 83–84
  credit spread risks, 7–9
  credit switches, 84
  reduced-credit-exposure uses, 83–84
  returns-enhancement uses, 83
  TRSs, 77–78
  zero-cost credit exposures, 84
Predominately-speculative-grade credit ratings, 8t
Present values (PVs)
  CDSs, 101–104
  concepts, 101–104, 146
Prices, 2, 33–34, 71
  approaches, 88–92

asset swaps, 88, 96–98, 123–127
Black–Scholes–Merton model, 88, 93, 94t, 95t
bond/equity derivatives, 33–34
CDSs, 71, 90, 98, 101, 123
concepts, 2–3, 12–13, 17, 19, 22–23, 27–30, 32–34, 38, 42–43, 45–46, 53, 57, 66–67, 69–71, 76–79, 101, 123
credit derivatives, 2, 33–34, 87–92, 101, 123
credit spread options, 102
credit spreads, 92–94
Das–Tufano model, 90
Duffie–Singleton model, 91–92
Hull–White model, 102
Jarrow, Lando and Turnbull model, 89–90
JPMorgan approach, 47, 134
market approach, 104–106
mean-reversion model prices, 93–94
models, 88–92, 124
no-arbitrage pricing models, 89, 104, 110, 117, 122, 124
recovery rates, 92, 98–99, 102
reduced-form models, 89–92, 97–98, 102
structural models, 88
theoretical pricing approach, 101–104
TRSs, 33, 76
Probability of default
price factors, 88–92
see also Credit ratings
Prospectuses, 11
PVs see Present values

R

Recovery rates
concepts, 12–15
prices, 88–92, 98–99, 102
seniority levels, 12–13, 103
Redemption values, 22, 53
Reduced-form pricing models
concepts, 88–92
key features, 89
liquidity issues, 89
Reference assets/entities
concepts, 27, 42–43, 62–64, 64f, 83, 123–125
market values, 40–41
Relative value trading, 71–75
example, 73–75
rationale, 73

Repo transactions
TRSs, 32, 78
Restructuring events, 7, 18–21, 44–45, 49–50
Reuters, 33, 40
Risk-free bonds, 56, 89, 91, 95–96, 101–102, 126f
Risk-neutral concepts see No-arbitrage...
Risks
bonds, 56, 89, 91, 95–96, 101–102, 126f
cash/synthetic position comparisons, 45
CDSs, 44–45, 83–84, 98–100, 125–127
CLNs, 44–45, 64, 83–84
credit derivatives, 17–21, 44–45, 65–67, 123–127
credit-event definitions, 44–45
interest rates, 19, 22–23, 30–32, 68, 77
TRSs, 30–32, 77
types, 4–12, 44–45, 98
see also Credit...; Market...
'Risky curves', CDSs, 106

S

S&P see Standard & Poor's (S&P)
Schwartz, E.S., 102
Seniority levels, recovery rates, 12–13
Settlement agreements, 20–22, 24, 40–44, 41f, 54, 63
concepts, 20–22, 24, 40–44, 41f, 54, 63
market requirements, 42–43
mechanisms, 41–44
Singleton, K., 89, 91–92
Special bonds, 77, 126, 147
Special purpose legal entities (SPEs), 54, 61–62, 143
Special purpose vehicles (SPVs), 54, 61–62, 143
Speculation, 8t
Speculative-grade credit ratings, 8t
SPEs see Special purpose legal entities
SPVs see Special purpose vehicles
Standard & Poor's (S&P), 7, 8t, 11–12, 20
Stochastic recovery rates, 90
Structural pricing models, concepts, 88
Structured Credit Products: Credit Derivatives and Synthetic Securities (author), 6
Structured products, 27, 40, 42, 54–55, 61–64, 83, 108
CLNs, 27, 40, 42, 54–55, 61–64, 83
Synthetic repo transactions, 32, 34, 75–78
Synthetic structured products, 30–32, 45, 65, 83
Systemic risks, 17, 98

T

T-bills, 61–62, 145
Technical default, concepts, 7, 10, 20, 26, 126
Term structure
  of credit rates, 19, 111, 121f, 122, 132
  of default probabilities, 117, 121f, 122
  of interest rates, 106
  of risky bonds, pricing models, 19, 38–39, 89–90, 92
  of spreads by maturity, 93, 98
Thailand, 6
Time value for money, definition, 145–147
Tolk, J., 44
Total return swaps (TRSs), 30–36, 75–79, 98
  applications, 75–79
  balance sheets, 75–76, 78–79
  capital structure arbitrage, 76–77
  concepts, 30–36, 75–79, 98
  definitions, 30
  funded forms, 33, 34b
  generic details, 30–36
  maturity issues, 30–32, 75–79
  prices, 33, 75–79, 90, 98
  repo, 77–78
  risks, 34, 77
Transition matrices, credit ratings, 89–90
Transparency benefits, credit derivatives, 17
TRSs *see* Total return swaps
Tufano, P., 89

U

Underlying, definition, 147
Underwriting, definition, 147

Unfunded credit derivatives
  concepts, 17
  definition, 22–23
  see also Credit default swaps
Unintended risks, CDSs, 44
US
  credit rating agencies, 11–12
  restructuring events, 20–21

V

Valuations, 87, 101
  concepts, 87
  rule-of-thumb, 106
  see also Prices
Value-at-risk measurement, credit risks, 147
Vanilla credit options, 102–103
Verification agents, credit events, 40
Volatility
  concepts, 6, 34–36, 73, 89–90, 92–93, 98–99, 106–108, 145, 148
  credit curves, 99
  credit spread, 93

W

White, A., 102

Y

Yield
  credit spread risks, 7–9, 94–96
  credit-risk links, 6, 9f, 10, 30
  definitions, 148
  forward credit spread, 94–96

Z

Zero-cost credit exposures, portfolios, 84

CPI Antony Rowe
Eastbourne, UK
June 20, 2023